计算机系列教材

汤晓兵　徐遵义　主编
赵洪銮　白彧　袁卫华　副主编

计算机新技术教程

清华大学出版社
北京

内 容 简 介

本书精选和提炼了目前计算机专业领域最新的研究成果和学术进展，内容取材广泛，难易适中，每单元各具特色。通过对文章中专业词汇、语法结构的介绍和分析，为读者将来在信息技术以及相关研究方向的英文资料的阅读、翻译和写作等能力提供了重要的支撑和保证。本书结合作者多年计算机教学和科研中的宝贵经验，以及与国外学术交流的体会总结，在保证本书专业知识领先的同时，更着重体现了内容的实用性和针对性。读者在提高专业英语读、写、译水平的同时，可通过本书及时地跟踪 IT 领域内的顶尖技术，把握学术上的研究热点，确立未来的学习和科研方向。

本书面向的读者是高校信息技术相关专业的本科生、大专生或是从事软、硬件开发及相关领域的工程技术人员，通过对每个单元中英文文献专题的引入、分析和解决问题的过程，培养和提高读者在专业英语方面的阅读、翻译和写作的综合能力。

本书封面贴有清华大学出版社防伪标签，无标签者不得销售。
版权所有，侵权必究。举报：010-62782989，beiqinquan@tup.tsinghua.edu.cn。

图书在版编目(CIP)数据

计算机新技术教程：英文/汤晓兵，徐遵义主编. —北京：清华大学出版社，2016（2024.7重印）
计算机系列教材
ISBN 978-7-302-43410-8

Ⅰ.①计… Ⅱ.①汤… ②徐… Ⅲ.①计算机技术—教材—英文 Ⅳ.①TP3

中国版本图书馆 CIP 数据核字(2016)第 075218 号

责任编辑：白立军　李　晔
封面设计：常雪影
责任校对：时翠兰
责任印制：刘海龙

出版发行：清华大学出版社
网　　址：https://www.tup.com.cn，https://www.wqxuetang.com
地　　址：北京清华大学学研大厦 A 座　　邮　编：100084
社 总 机：010-83470000　　　　　　　　　邮　购：010-62786544
投稿与读者服务：010-62776969，c-service@tup.tsinghua.edu.cn
质量反馈：010-62772015，zhiliang@tup.tsinghua.edu.cn
课件下载：https://www.tup.com.cn，010-83470236

印 装 者：三河市龙大印装有限公司
经　　销：全国新华书店
开　　本：185mm×260mm　　印　张：13　　字　数：299 千字
版　　次：2016 年 8 月第 1 版　　　　　　 印　次：2024 年 7 月第 5 次印刷
定　　价：39.00 元

产品编号：068059-02

《计算机新技术教程》 前言

本书在学习目标上有两个方面：一是从语言的角度来进行文章中英语的学习；二是对计算机专业领域相关知识的介绍，后者涵盖了从计算机的基本硬件和软件系统组成到计算机发展的前沿技术。在具体实施中，通过对这两个方面的交互渗透，在学好英语语义、语法的同时，让读者学习不同研究方向的计算机专业文献，对知识进行综合的学习和掌握，培养和提高读者在英文阅读、写作和翻译三个方面的专业英语综合学习和工作能力。

本书共分12个单元：第1单元硬件基础（Hardware），第2单元软件基础（Software），第3单元数据库（Database）为汤晓兵副教授编写，第4单元程序设计（Programming），第5单元办公计算（Office Computing），第6单元计算机网络（Networking）为徐遵义副教授编写，第7单元移动设备计算（Mobile Application）为赵洪銮副教授编写，第8单元网络开发（Web Development）为袁卫华讲师编写，第9单元计算机安全（Security）为王庆东高级工程师编写，第10单元网络服务（Web Services）为詹玲讲师编写，第11单元大数据（Big Data）为白彧助理研究员编写，第12单元云计算（Cloud Computing）为康喆讲师编写。汤晓兵、袁卫华和徐晓静编写了每单元后针对词汇、语法结构和单元综述的测试与练习。汤晓兵负责全书的统稿。

在本书的编写过程中，得到了清华大学出版社的编辑以及刘祥昭老师的大力支持和帮助，在此对他们表示由衷的感谢。同时，感谢我的父母、妻子和双胞胎儿子们。在我编辑忙碌的时候，是他们的关心和爱护，支持着我集中精力、全力以赴地把书稿顺利完成。

由于编者水平有限，书中的缺点和不足难免，恳请读者指正。

编 者

Unit 1　Hardware—Microwave Integrated Circuits　/1
　1.1　Classification of Microwave Integrated Circuits　/1
　1.2　Microwave Circuits in a Communication System　/6
　1.3　Summary　/9
　1.4　New Words and Expressions　/10
　1.5　Questions　/10
　1.6　Problems　/12

Unit 2　Software—Microsoft Project 2013 and the Project Management Domain　/13
　2.1　History of Project Management　/13
　2.2　Exploring Project Management Industry Standards　/14
　　2.2.1　Project Management Body of Knowledge (PMBOK)　/14
　　2.2.2　PRINCE2　/18
　2.3　WBS, Phases and Control Points, Methodologies, and Life Cycles　/20
　　2.3.1　Work Breakdown Structure(WBS)　/20
　　2.3.2　Managerial Control　/21
　　2.3.3　Phases and Gates　/21
　　2.3.4　Methodologies　/22
　　2.3.5　Life Cycles　/22
　2.4　Using Microsoft Project with Methodologies and Life Cycles　/22
　　2.4.1　Waterfall Development Process　/23
　　2.4.2　Iterative Development　/23
　　2.4.3　Research Project　/27

2.5　Accommodating Teaming Styles　/27
2.6　Consultants' Tips　/28
　　2.6.1　Determine the Approach to Use in Managing Your Project　/28
　　2.6.2　Use WBS as a First Step in Project Definition　/28
　　2.6.3　Use the 5×9 Checklist for Planning　/29
2.7　New Words and Expressions　/29
2.8　Questions　/30
2.9　Problems　/31

Unit 3　Database—NoSQL Databases: An Overview　/32
3.1　NoSQL: What does it mean　/32
3.2　Why NoSQL Databases　/32
3.3　Aggregate Data Models　/33
3.4　Distribution Models　/34
3.5　CAP theorem　/35
3.6　Types of NoSQL Databases　/36
　　3.6.1　Key-Value databases　/36
　　3.6.2　Document databases　/37
　　3.6.3　Column family stores　/38
　　3.6.4　Graph Databases　/38
3.7　Why choose NoSQL database　/40
3.8　Choosing NoSQL database　/41
3.9　Schema-less ramifications　/41
3.10　Conclusion　/42
3.11　New Words and Expressions　/42
3.12　Questions　/42
3.13　Problems　/45

Unit 4　Programming—What Software Architects Need to Know About DevOps　/46
　4.1　What Software Architects Need to Know About DevOps　/46
　4.2　Defining DevOps　/46
　4.3　DevOps Practices and Architectural Implications　/48
　4.4　Organizational Aspects of DevOps　/50
　4.5　Implications for Software Architecture: Microservices　/51
　4.6　Summary　/52
　4.7　New Words and Expressions　/53
　4.8　Questions　/53
　4.9　Problems　/56

Unit 5　Office Computing—The Ultimate Player's Guide to Minecraft-Xbox Edition: Gathering Resources　/57
　5.1　Introducing the HUD　/57
　5.2　Avoiding Getting Lost　/60
　5.3　Improving Your Tools　/61
　5.4　Chests: Safely Stashing Your Stuff　/63
　5.5　Avoiding Monsters　/64
　5.6　Hunger Management　/66
　5.7　Your Mission: Food, Resources, and Reconnaissance　/67
　　5.7.1　Food on the Run　/68
　　5.7.2　Finding a Building Site　/70
　5.8　A Resourceful Guide to the Creative Mode Inventory　/72
　5.9　The Bottom Line　/74
　5.10　New Words and Expressions　/74
　5.11　Questions　/75
　5.12　Problems　/77

Unit 6　Networking—Troubleshooting Methods for Cisco IP Networks　/78

 6.1　Troubleshooting Principles　/78

 6.2　Structured Troubleshooting Approaches　/81

  6.2.1　The Top-Down Troubleshooting Approach　/83

  6.2.2　The Bottom-Up Troubleshooting Approach　/84

  6.2.3　The Divide-and-Conquer Troubleshooting Approach　/85

  6.2.4　The Follow-the-Path Troubleshooting Approach　/86

  6.2.5　The Compare-Configurations Troubleshooting Approach　/87

  6.2.6　The Swap-Components Troubleshooting Approach　/88

 6.3　Troubleshooting Example Using Six Different Approaches　/89

 6.4　Summary　/91

 6.5　New Words and Expressions　/91

 6.6　Questions　/92

 6.7　Problems　/94

Unit 7　Mobile Application—What's Special about Mobile Testing?　/95

 7.1　User Expectations　/96

 7.2　Mobility and Data Networks　/97

 7.3　Mobile Devices　/98

 7.4　Mobile Release Cycles　/100

 7.5　Mobile Testing Is Software Testing　/101

 7.6　Summary　/102

 7.7　New Words and Expressions　/102

7.8　Questions　/103
7.9　Problems　/105

Unit 8　Web Development—The Mobile Commerce Revolution and the Current State of Mobile　/106
8.1　Americans and Smartphones　/106
8.2　Mobile Around the World　/109
8.3　Mobile Commerce　/110
8.4　Beyond the Numbers　/110
8.5　The Bottom Line　/113
8.6　New Words and Expressions　/113
8.7　Questions　/114
8.8　Problems　/116

Unit 9　Security—Information Security Principles of Success　/117
9.1　Introduction　/117
9.2　Principle 1: There Is No Such Thing As Absolute Security　/117
9.3　Principle 2: The Three Security Goals Are Confidentiality, Integrity, and Availability　/119
　　9.3.1　Integrity Models　/119
　　9.3.2　Availability Models　/120
9.4　Principle 3: Defense in Depth as Strategy　/120
9.5　Principle 4: When Left on Their Own, People Tend to Make the Worst Security Decisions　/122
9.6　Principle 5: Computer Security Depends on Two Types of Requirements: Functional and Assurance　/122
9.7　Principle 6: Security Through Obscurity Is Not an Answer　/123
9.8　Principle 7: Security = Risk Management　/124

9.9 Principle 8: The Three Types of Security Controls Are Preventative, Detective, and Responsive /126

9.10 Principle 9: Complexity Is the Enemy of Security /127

9.11 Principle 10: Fear, Uncertainty, and Doubt Do Not Work in Selling Security /127

9.12 Principle 11: People, Process, and Technology Are All Needed to Adequately Secure a System or Facility /127

9.13 Principle 12: Open Disclosure of Vulnerabilities Is Good for Security! /128

9.14 Summary /129

9.15 New Words and Expressions /129

9.16 Questions /130

9.17 Problems /131

Unit 10 Web Services—Designing Software in a Distributed World /132

10.1 Visibility at Scale /133

10.2 The Importance of Simplicity /134

10.3 Composition /134

 10.3.1 Load Balancer with Multiple Backend Replicas /134

 10.3.2 Server with Multiple Backends /136

 10.3.3 Server Tree /138

10.4 Distributed State /139

10.5 The CAP Principle /142

 10.5.1 Consistency /142

 10.5.2 Availability /143

 10.5.3 Partition Tolerance /143

10.6 Loosely Coupled Systems /145

10.7 Speed /147

10.8 Summary /150

10.9 New Words and Expressions /151

10.10 Questions /151

10.11 Problems /153

Unit 11 Big Data—Big Data Computing /154

11.1 Introduction /154

11.2 Apache Hadoop Data modelling /156

11.3 NoSQL Big Data systems /157

 11.3.1 Key Value stores /157

 11.3.2 Document databases /157

 11.3.3 Graph databases /158

 11.3.4 XML databases /158

 11.3.5 Distributed Peer Stores /158

 11.3.6 Object stores /159

11.4 Definitions of Data Management /159

 11.4.1 Data management /159

 11.4.2 Big data management (BDM) /160

11.5 The State of Big Data Management /160

11.6 Big Data Tools /162

 11.6.1 Big data tools: Jaspersoft BI Suite /162

 11.6.2 Big data tools: Pentaho Business Analytics /163

 11.6.3 Big data tools: Karmasphere Studio and Analyst /163

 11.6.4 Big data tools: Talend Open Studio /164

 11.6.5 Big data tools: Skytree Server /164

 11.6.6 Big data tools: Tableau Desktop and Server /165

 11.6.7 Big data tools: Splunk /165

11.7 New Words and Expressions /166

11.8　Questions　/166

11.9　Problems　/168

Unit 12　Cloud Computing—The Practice of Cloud System Administration: Operations in a Distributed World　/169

12.1　Distributed Systems Operations　/170

　　12.1.1　SRE versus Traditional Enterprise IT　/170

　　12.1.2　Change versus Stability　/171

　　12.1.3　Defining SRE　/173

　　12.1.4　Operations at Scale　/174

12.2　Service Life Cycle　/177

　　12.2.1　Service Launches　/178

　　12.2.2　Service Decommissioning　/180

12.3　Organizing Strategy for Operational Teams　/180

　　12.3.1　Team Member Day Types　/183

　　12.3.2　Other Strategies　/185

12.4　Virtual Office　/186

　　12.4.1　Communication Mechanisms　/187

　　12.4.2　Communication Policies　/187

12.5　Summary　/188

12.6　New Words and Expressions　/189

12.7　Questions　/190

12.8　Problems　/191

Answers to Questions　/193

References　/194

Unit 1　Hardware—Microwave Integrated Circuits

This chapter covers the following topics:
- 1.1　Classification of Microwave Integrated Circuits
- 1.2　Microwave Circuits in a Communication System
- 1.3　Summary

1.1　Classification of Microwave Integrated Circuits

An active microwave circuit can be defined as a circuit in which active and passive microwave devices such as resistors, capacitors, and inductors are interconnected by transmission lines. At low frequencies, the transmission lines are a simple connection; however, at microwave frequencies they are no longer just simple connections and their operation becomes a complicated distributed circuit element. As a result, a microwave integrated circuit's classification is based on the fabrication method of the transmission lines used for interconnection.

There are various types of transmission lines in microwave integrated circuits; some common examples are waveguides, coaxial, and microstrip lines. Figure 1.1 shows the transmission lines used in microwave circuits. Although there are special cases of microwave integrated circuits that are composed of coaxial lines and waveguides, in most cases the microwave integrated circuits are formed using planar transmission lines. Therefore, the content of this book is restricted to microwave integrated circuits formed using planar transmission lines, examples of which are microstrip, slot line, and co-planar waveguide (CPW), as shown in Figure 1.2. These planar transmission lines are frequently used in the large-scale production of microwave circuits and generally form the basic transmission lines for microwave circuits.

The implementation of planar transmission lines on substrates can be classified into two basic groups: *monolithic* and *hybrid integrated circuits*. In monolithic integration, the active and passive devices as well as the planar transmission lines are grown *in situ* on one planar substrate that is usually made from a semiconductor material called a wafer.

Figure 1.3 shows an example of monolithic integration. Figure 1.3(a) is a photograph of the top side of a wafer and Figure 1.3(b) shows a single monolithic

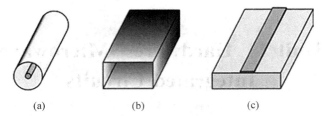

Figure 1.1 Some common transmission lines used in microwave circuits: (a) coaxial line, (b) rectangular waveguide, and (c) microstrip line.

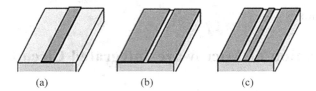

Figure 1.2 Some common planar transmission lines used in microwave circuits: (a) microstrip, (b) slot line, and (c) CPW (co-planar waveguide).

microwave integrated circuit; the identical circuits are repeatedly produced on the wafer in Figure 1.3(a). The monolithic microwave integrated circuit in Figure 1.3(b) is found to contain active and passive devices, and planar transmission lines. The monolithic integration provides a compact sized circuit and eliminates a significant amount of assembly when building a component or a system. Especially because size is of critical importance in most recent RF systems, monolithic integration is frequently employed to provide a compact component. An advantage of monolithic integration is that it is well suited for large-scale production, which results in lower costs. A disadvantage is that monolithic integration takes a long time to develop and fabricate, and small-scale production results in highly prohibitive costs.

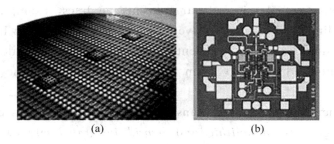

Figure 1.3 Monolithic integration: (a) a wafer and (b) a monolithic microwave integrated circuit on the wafer.

Hybrid integration is a fabrication method in which the transmission lines are implemented by conductor patterns on a selected substrate with either *printing* or *etching*, and active and passive devices are assembled on the patterned substrate by either soldering or wire bonding. When implementing transmission lines by conductor patterns on a substrate, careful consideration must be given to the substrate material and the conductor material for the transmission lines because these materials can have significant effects on the characteristics of transmission lines. Hybrid integration is thus classified into three types based on the method by which the lines are formed on the substrate: a *printed circuit board* (PCB), a *thick-film* substrate, and a *thin-film* substrate.

Figure 1.4 shows an example of how connection lines are formed on a PCB substrate. Both sides of the dielectric material are attached with copper cladding that is then etched to obtain the desired conductor patterns. For PCB substrate materials, epoxy fiberglass (FR4), teflon, and duroid are widely used. FR4 substrate (a kind of epoxy fiberglass) can be used from lower frequencies to approximately 4 GHz, while teflon or duroid can be used up to the millimeter wave frequencies, depending on their formation. Generally, all these materials lend themselves to soldering while wire bonding for an integrated circuit assembly is typically not widely used. Furthermore, compared with other methods that will be explained later, a PCB can result in lower costs; its fabrication is easy and requires less time to produce. In addition, production on a small scale is possible without the use of expensive assembly machines; it is easy to fix and could also be used in large-scale production, and is thus widely used.

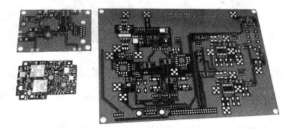

Figure 1.4　A photograph of epoxy fiberglass PCBs.

The PCBs on the left are for the X-band and 2 GHz frequency synthesizers using the phase locked loop. The PCB on the right is for the VHF automatic identification system, which has a similar block diagram shown in Figure 1.7. The power amplifier is implemented in a separate block.

Thick-film substrates are produced by screen-printing techniques in which conductor patterns are formed by pushing conductive paste on a ceramic substrate through a patterned screen and then firing printed conductor patterns. The substrate is called thick film because the patterns formed by such techniques are generally much thicker than those formed using thin-film techniques. As a benefit of using screen-printing techniques, multiple screen printings are possible. Dielectric or resistor patterns can also be formed by similar screen-printing techniques using dielectric or resistor pastes. Using an appropriate order of multiple screen printings, it is also possible to form capacitors and resistors on the ceramic substrate. Since the ceramic substrate is more tolerant of heat, it is easy to assemble active devices in the form of chips. On the other hand, considering the lines and patterns formed by this process, the pattern accuracy of thick film is somewhat inferior compared to that of thin film. The costs and development time, on a case-by-case basis, are somewhere between those of the PCB and thin-film processes. Recently, however, the integration based on thick-film technology has become rare because its cost and pattern accuracy are between the PCB and thin-film technology, while thick film is widely used to build multifunction components. A typical example is the package based on LTCC (low-temperature co-fired ceramics) technology Figure 1.5 Multilayer ceramics and structuring are possible in LTCC technologies. Figure 1.5 shows a photograph of thick-film patterned substrates fabricated using the thick-film process.

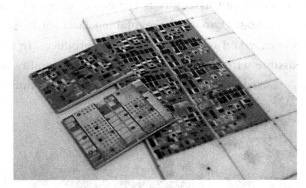

Figure 1.5 A photograph of substrates fabricated by the thick-film process.

The thin-film technique is very widely used in the fabrication of microwave circuits for military and microwave communication systems. In the case of the thin-film process, a similar ceramic substrate material used in thick film is employed, but compared to the thick-film substrate, a fine surface-finish substrate is used. The most widely used substrate is 99% alumina (Al_2O_3). Other substrates such as fused silica, quartz, and so on are possible for conductor-pattern generation based on thin-film technologies. The pattern formation on the substrate is created with a photolithographic process that can produce fine tracks of conductor patterns similar to those in a semiconductor process. Since the thin-film substrate is also alumina as in the case of a thick-film substrate, the assembly of semiconductor chips using wire bonding is possible. Thin film compared with PCB and thick film is more expensive, and due to the requirement of fine tracks, a mask fabrication is necessary and the process generally takes longer. Passive components such as resistors and air-bridge capacitors can be implemented using this process. In addition, integrated circuits produced by the thin-film process require special wire bonders and microwelding equipment for assembly. Compared to the monolithic integration process, the thin-film process tends to be cheaper in terms of cost, but compared to MMIC, the assembled circuit using the thin-film patterned substrate is difficult to characterize precisely because of unknown or poorly described parasitic circuit elements associated with the assembly methods such as wire bonding and die attach. Before the emergence of MMICs (monolithic microwave integrated circuits), thin-film technology was the conventional method for building microwave-integrated circuits (MICs). Figure 1.6 is a photograph of thin-film circuits fabricated with the thin-film technique.

Figure 1.6 A photograph of substrates produced by the thin-film process.

From top left to bottom right, they are filter, phase shifter, power amplifier, path-switching circuit by assembly, power divider, and 50 Ω lines.

The choice of integration method depends on the application and situation, taking into account several factors mentioned previously, such as the operating frequency of the integrated circuit, the types of semiconductor components (chip or packaged), the forms of the passive components, large-scale fabrication costs, and method of assembly. These factors should all be considered when selecting the optimum method of integration. Table 1.1 provides a comparison of the hybrid integrations described previously.

Table 1.1 Comparison of hybrid integration

Technology	Cost	Fabrication Time	Pattern Accuracy	Assembly
PCB	Low	Short	Low	Soldering
Thick film	Middle	Middle	Low	Soldering and wire bonding
Thin film	High	Long	Fine	Soldering and wire bonding

Now we will consider the application of the planar transmission lines such as microstrip, slot, and CPW to the monolithic and hybrid integration technologies. Microstrip lines are the most widely used transmission lines for both monolithic and hybrid integration technologies. In microstrip lines, the top conductor pattern is usually connected to the ground by a through hole or a via hole. Thus, the back-side process for the through-hole or via-hole fabrication is essential to building a circuit based on microstrip lines. This backside process is inconvenient especially in the monolithic integration. In hybrid integration, the holes can be fabricated through simple mechanical drilling for a PCB case and through laser or ultrasonic drilling for thick- and thin-film cases. Then, plating the fabricated holes completes the fabrication of a through or via hole. However, to fabricate via holes in monolithic integration, a wafer that typically has a normal thickness of about 600 mm should be polished down to about 100 mm thickness. Current technology does not support via-hole fabrication beyond 100 mm. In Figure 1.2, we can see that the CPW and slot lines do not need the back-side metallic ground and they eliminate the need for any additional backside metallization process. The CPW is very helpful in monolithic integration and is widely used to build MMICs without vias. However, the discontinuities of CPWs are not well understood compared to those of microstrip lines and the integration based on a CPW is not as popular as that based on a microstrip. The various discontinuities of microstrip and slot lines, CPWs, and planar transmission lines are covered in reference 2 at the end of this chapter.

1.2 Microwave Circuits in a Communication System

Microwave integrated circuit classification has been discussed previously. The microwave integrated circuit was classified according to the method of implementing the planar transmission lines for the purpose of connecting active and passive devices. The functions of microwave integrated circuits vary greatly and we will now consider several important microwave integrated circuits. Some examples of these circuits are low-noise amplifiers (LNA), power amplifiers (PA), oscillators, mixers, directional couplers, switches, attenuators, and filters, among a host of other microwave-integrated circuits. Among these, directional couplers, switches, attenuators, filters, and so on, are basically passive microwave circuits, although they are very widely used. Thus, they are not covered in this book because they are considered to be outside its scope. In addition, although components such as switches, variable attenuators, phase shifters, and other control circuits are important and are composed of semiconductor devices, they are generally not regarded as the basic building blocks of a wireless communication system. Therefore, this book will only cover low-noise amplifiers, power amplifiers, oscillators, and mixers, which are the most widely used circuits in the construction of wireless communication systems. The basic design theory of these circuits as well as the devices related to them will be explained in this book.

As an example of a wireless communication system, Figure 1.7 shows a block diagram of an analog cellular phone handset (Rx frequency is 869—894 MHz and Tx frequency is 824—849 MHz). A general transceiver used for the transmission and reception of analog signals (usually voice) has a similar block diagram that is shown in Figure 1.7. A weak RF signal with a typical power level of about -100 dBm (0.1 nW) received from an antenna first goes through a filter called a diplexer and the signal is received only in the receiver frequency band. The filtered signal is too weak for direct demodulation or signal processing, and a low-noise amplifier (LNA) with a gain of 20—30 dB is required to amplify the received signal. Too much gain may cause distortion and an LNA with a gain of 20—30 dB is usually employed.

Next, because the received signal frequency is so high, the first mixer shown in Figure 1.7 translates the carrier frequency to a lower frequency band called first IF (intermediate frequency). A double-conversion super heterodyne receiver is more widely used than a single-conversion super heterodyne receiver in a communication system. The filter in front of the first mixer again suppresses both the image frequency signal and other signals at the outside of the receiving frequency band. Since multiple users in service are using the same frequency band, multiples of other user signals generally coexist with the signal in the first IF. Intermodulations among the multiple signals are one of the crucial issues in mixer design. In order to filter out possible spurious signals that appear at the first mixer output, the signal is passed through a narrow bandpass filter that has a bandwidth of about the signal bandwidth. The first IF filter removes many unwanted spurious signals although it may not be completely sufficient. The first IF output is converted again through the second mixing. Now the center frequency of the second IF is low enough, the highly selective filter is available, and the spurious signals can be sufficiently suppressed through the second IF filter. In addition, the signal frequency is low enough and can be demodulated for the recovery of the original signal. The demodulator is an FM demodulator and is almost the same as the FM demodulator that is commercially popular.

Tx_EN stands for Tx enable and ALC stands for automatic level control. Tx_ and Rx_ data are required to set the programmable frequency dividers in Tx and Rx synthesizers. LE stands for Load Enable. When LE is high, the digital channel data are loaded to the corresponding programmable frequency divider in PLL IC. Lock signal indicates that the synthesizer using PLL is in a locked state.

Note that the mixer requires the input signal from a *local oscillator* (LO) for the

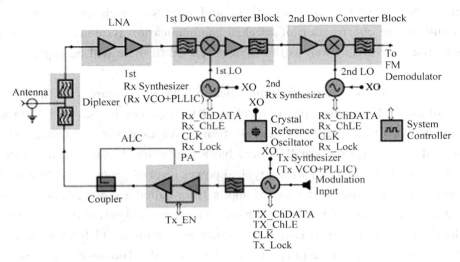

Figure 1.7 A block diagram of an analog mobile phone handset (AMPS standard).

translation of the signal frequency to the IF. The two LO signals are supplied from the two Rx-synthesizers and each Rx-synthesizer consists of a voltage-controlled oscillator (VCO) and a commercial PLL (phase-locked loop) IC (integrated circuit). Since the frequency of most VCOs is not stable enough to be used in such communication systems, the frequency of a VCO must be stabilized using a stable crystal oscillator (XO in Figure 1.7) with a typical temperature stability of 2 ppm (parts per million) and a phase-locked loop (PLL). Furthermore, the LO frequency should be moved up and down according to the base station commands. Such frequency synthesis and stabilization can be achieved by a phase-locked loop (PLL). To build a frequency synthesizer using PLL, the VCO frequency as well as the crystal oscillator frequency must be divided by appropriate programmable frequency dividers in the PLL IC. The signals CLK, Rx_ChDATA, Rx_ChLE, and Rx_Lock, shown in Figure 1.7, are the digital signals between the PLL IC and the system controller. The clock signal CLK is used for the timing reference signal that is generated by the system controller using the crystal oscillator. Rx_ChDATA sent from the controller represents the digital data to set the programmable frequency dividers. The signal Rx_ChLE selects the corresponding programmable divider for Rx_ChDATA to be loaded among several frequency dividers in the PLL IC. When phase lock is achieved, the PLL IC sends the signal Rx_Lock to the system controller to inform the phase lock completion. The two Rx synthesizers are necessary for the double-conversion superheterodyne receiver. The commercial PLL IC generally includes the necessary components to achieve the phase lock for two VCOs in a single PLL IC. Thus, the LO signal for the second conversion is similarly synthesized using a single PLL IC.

In the transmission operation, the modulation input signal (usually voice) goes to the modulation input of a Tx synthesizer. The Tx synthesizer is similarly composed of a VCO and a PLL IC. Through the PLL IC, the desired carrier center frequency is similarly synthesized as in the Rx synthesizer. The digital signals CLK, Tx_ChDATA, Tx_ChLE, and Tx_Lock are similarly interpreted as in the Rx synthesizer. The modulation signal has a generally higher frequency than the PLL loop bandwidth and thus can modulate a VCO without the effects of a PLL. Therefore, the *frequency-modulated* (FM) signal appears at the Tx synthesizer output with the synthesized carrier frequency. The modulated signal then passes through the bandpass filter that removes unnecessary or spurious signals. The average output power level of the modulated signal is generally low; thus, in order to obtain the desired RF power output level, the signal must be amplified by a power amplifier (PA) whose typical maximum output power level is about 1 W. The function ALC (Automatic Level Control) is generally built in to control the transmitting power level. When a user is close to the base station, the transmitting power level is set to low; otherwise, it is set to high for a better quality of communication. The PA output signal is then passed through a diplexer without affecting the receiver and radiated via the antenna. A power amplifier is important in this type of communication system because it consumes most of the DC power supplied from a battery. Furthermore, because a power amplifier operates in large-signal conditions, significant distortion arises.

Given the preceding discussion, the key circuits in building a communication system are a low-noise amplifier, a power amplifier, oscillators, and mixers. With that in mind, this book will discuss in detail the design and evaluation method of these circuits.

1.3 Summary

- Microwave integrated circuits can be classified according to the fabrication method of the patterned substrate and in terms of monolithic and hybrid integration. Hybrid integration can be further classified into integrations based on PCB, thick film, and thin film. In the selection of integration, one type cannot be said to be superior to the other; the choice is made depending on the application and given situation, and by taking into consideration several factors such as cost, time, pattern accuracy, and assembly.
- Among active microwave circuits, the most commonly used building blocks for wireless communication systems or other systems, such as repeaters, transponders, and radars, are amplifiers, oscillators, and mixers.

1.4 New Words and Expressions

integrated circuits	n. 集成电路
coaxial line	n. 同轴电缆
monolithic integration	n. 单片集成电路
hybrid integrated circuit	n. 混合集合电路
semiconductor material	n. 半导体材料
fabricate	vt. 制造；伪造；装配
etch	v. 蚀刻
PCB (printed circuit board)	abbr. 印刷电路板
epoxy fiberglass	n. 环氧树脂玻璃纤维
teflon	n. 聚四氟乙烯
duroid	n. 杜罗艾德铬合金钢
soldering	v. 焊接；修补
frequency synthesizer	n. 频率合成器
VHF (Very High Frequency)	abbr. 甚高频
ceramic substrate	n. 陶瓷基片；陶瓷衬底
case-by-case	具体分析
LTCC (Low Temperature Co-Fired Ceramic)	abbr. 低温共烧陶瓷
photolithographic process	n. 光刻工艺；制版工艺
passive components	n. 无源元件
in terms of	依据；按照
MMIC (Microwave Monolithic Integrated Circuit)	abbr. 微波单片集成电路
die attach	n. 管芯连接
taking into account	顾及，考虑
directional coupler	n. 定向耦合器
VCO (voltage-controlled oscillator)	abbr. 电压控制振荡器

1.5 Questions

Single Choice Questions

1. Which is not the type of transmission lines in microwave integrated circuits(　　)?
   　　A. waveguides　　　　　　　　B. coaxial
   　　C. microstrip lines　　　　　　D. resistor

2. The implementation of planar transmission lines on substrates can be classified into two basic groups：(　　).
   　　A. monolithic and hybrid integrated circuits

B. circuits and monolithic

C. monolithic and lines

D. hybrid integrated circuit and lines

3. The monolithic microwave integrated circuit in Figure 1.3(b) is found to contain(   ).

   A. active and passive devices and planar transmission lines

   B. active and passive devices

   C. active devices and planar transmission lines

   D. passive devices and planar transmission lines

4. Hybrid integration is thus classified into three types based on the method by which the lines are formed on the substrate: (   ).

   A. conductor, a thick-film substrate, and a thin-film substrate

   B. a printed circuit board (PCB), conductor, and a thin-film substrate

   C. a printed circuit board (PCB), a thick-film substrate, and a thin-film substrate

   D. a printed circuit board (PCB), a thick-film substrate, and conductor

5. Which PCB substrate materials is not widely used? (   )

   A. epoxy fiberglass   B. conductor

   C. teflon   D. duroid

6. Compared with other methods that will be explained later, (   ) can result in lower costs.

   A. conductor   B. a printed circuit board (PCB)

   C. a thick-film substrate   D. a thin-film substrate

7. What does the acronym LTCC stand for ? (   )

   A. lower-temperature co-fired ceramics

   B. low-temperature co-fired ceramic

   C. low-temperature co-fire ceramics

   D. low-temperature co-fired ceramics

8. Before the emergence of MMICs, (   ) was the conventional method for building microwave-integrated circuits.

   A. screen-printing technology

   B. printed circuit board (PCB) technology

   C. thick-film technology

   D. thin-film technology

9. Which are the most widely used circuits in the construction of wireless communication systems in this book? (   )

   A. power amplifiers, directional couplers, switches, and filters

   B. low-noise amplifiers, power amplifiers, oscillators, and mixers

        C. oscillators, directional couplers, switches, and attenuators

        D. mixers, switches, attenuators, and filters

10. Among active microwave circuits, the most commonly used building blocks for wireless communication systems or other systems, such as repeaters, transponders, and radars, are (　　).

        A. attenuators, oscillators and mixers

        B. amplifiers, attenuators and mixers

        C. amplifiers, oscillators and attenuators

        D. amplifiers, oscillators and mixers

1.6　Problems

After reading this chapter and completing the exercises, you will be able to do the following:

1. A waveguide generally has lower line loss than a microstrip. An SIW (substrate integrated waveguide) can be considered as the planar version of a waveguide. How is an SIW configured using a substrate?

2. Find the TR (transmission and receiving) module example built using a LTCC on the Web site www.barryind.com.

3. How is the ALC in Figure 1.7 constructed?

4. Refer to the FM demodulator IC SA605, which is used to demodulate an FM signal. Explain how the FM signal is demodulated using its block diagram.

5. Refer to the Web site of vendors of PLL IC such as Analog Devices Inc. or other companies. Explain the synthesizer data bus shown in Figure 1.7.

6. How can the PLL be modulated? Explain how to set the PLL loop bandwidth by taking the bandwidth of a bandlimited modulation signal into consideration.

Unit 2 Software—Microsoft Project 2013 and the Project Management Domain

2.1 History of Project Management

When Patrick Henry said, "I know of no way of judging the future but by the past," he could have been talking about project management. When faced with projects that have never been done before, all project managers can do is look at what has come before them.

Although project management has been practiced for thousands of years, evidenced by the Egyptian and Roman dynasties, modern project management can be traced back to the late nineteenth century and the rise in large-scale government projects and growing technological advancements. Fredrick Taylor, the Father of Scientific Management, applied scientific reasoning to analyzing and improving labor, and Henry Gantt studied management of Navy ship construction during World War I. Gantt's use of charts, task bars, and milestone markers is still practiced today, and they bear his name. One of the major projects that brought detailed project planning, controlling, and coordination to the forefront was the Hoover Dam project, which involved $175 million dollars, six different companies, a major worksite with no existing infrastructure, and approximately 5,200 workers. The project was brought in under budget and ahead of schedule.

After developments in project management during two World Wars and the growing Cold War, major changes to project management were brought about with the launch of Sputnik. Fearful that the United States was falling behind in the race to space, the United States introduced several major programs to focus on science and exploration. Several agencies, including the Advanced Research Project Agency, a high-level research and development program that later became DARPA, and NASA were founded. These agencies led the way in the development of project management.

Two other major developments for project management to grow out of this period were the Critical Path Method (CPM) and the Program Evaluation and Review Technique (PERT). CPM was devised by Du Pont and Remington Rand for use with the UNIVAC-1 computer mainframe. PERT was invented by the Program Evaluation Branch of the Special Projects office of the U.S. Navy, for use with the POLARIS missile program, and was also used on the Apollo program for NASA. CPM/PERT gave managers more control over extremely large and complex projects, but could only be calculated within large mainframe computer systems and were used mainly for government sector projects.

With the computer revolution of the 1980s and the move from mainframe computers to personal computers with the ability to multitask, project management software became more accessible to other companies. The Internet and networked systems only made project managers more efficient at controlling and managing the different aspects of their projects. More information on previously completed projects is available today than ever before, making the project manager's job of estimating the future by looking at the past easier than ever.

2.2 Exploring Project Management Industry Standards

Almost anyone can create a schedule with Project. Organizing that schedule into a logical flow of work, however, requires a solid understanding of how projects should be managed and decomposed into logical units. To understand project management, you must understand the standards and methodology behind it. Although Gantt Charts and other similar resources are used in almost all project management schedules, there are several different ways of using those resources.

This chapter discusses prominent industry standards often used to set a framework for building schedules. A variety of methodologies, team styles, and life cycles also are explored. The approach and techniques vary, but the software can still be used to support virtually any approach to scheduling that an individual or organization chooses to use.

2.2.1 Project Management Body of Knowledge (PMBOK)

The Project Management Institute, or PMI, is an internationally recognized organization that has developed standards for the domain of project management including standards for portfolio management, program management, project management, and Work Breakdown Structures. PMI has several hundred thousand members in more than 65 countries. It is widely recognized for its certification programs and continues to grow through a combination of volunteer efforts, certification programs, local chapter events, international seminars, and special interest groups.

The standards created by PMI are authored by a vast network of project management professionals who volunteer their time to create and update these standards on a regular basis. The standards groups are from many different countries across the globe; they research topics and collaborate to bring together the latest thinking and techniques from their collective experience.

The PMI standard that is of primary importance for this chapter of the book is in its fifth edition and is known as "A Guide to the Project Management Body of Knowledge," also known as the PMBOK Guide. It is discussed in some detail in this chapter to help in understanding all the components that should be considered when creating a schedule.

Because PMI is a standards and certification organization, it does not prescribe methodologies or "how to" approaches; rather, it defines specific standards and offers certifications in the field of project management. The PMBOK provides a context for a way to do things, rather than the process that should be followed.

Inexperienced project managers often try to make their schedules follow PMBOK as if it were a recipe for success. This can lead them into traps and complexity that is not useful in completion of their projects. Instead, they should look to the PMBOK for support of the methodology and life cycle that they choose to follow.

The PMBOK Guide has established five process groups to define the project management process. These processes are as follows:

- **Initiating Process Group**—Defines and authorizes the project or a project phase.
- **Planning Process Group**—Defines and refines objectives, and plans the course of action required to attain the objectives and scope that the project is to address.
- **Executing Process Group**—Integrates people and other resources to carry out the project management plan.
- **Monitoring and Controlling Process Group**—Regularly measures and monitors progress to identify variances from the project management plan so that corrective action can be taken when necessary to meet project objectives.
- **Closing Process Group**—Formalizes acceptance of the product, service, or result and brings the project or a project phase to an orderly end.

TIP

PMI makes a clear distinction between a *project plan* and a *project schedule*. The *plan* is a formal document that includes narrative on communication approaches, assumptions, deliverables, and execution of the project. The *schedule* is one component of the plan that focuses on the timeline for the activities to be performed. As scheduling tools become more sophisticated, they are gradually including more elements that used to reside only in the plan. Project Desktop still focuses on scheduling functions, but the server components have added capabilities to support more of the plan functions.

Projects are created and implemented in environments that are larger in scope than the projects themselves. All projects must have a beginning and an end, as shown by the Initiating and Closing process groups. In between, a project will be engaged continually with the other three process groups, as shown in Figure 2.1.

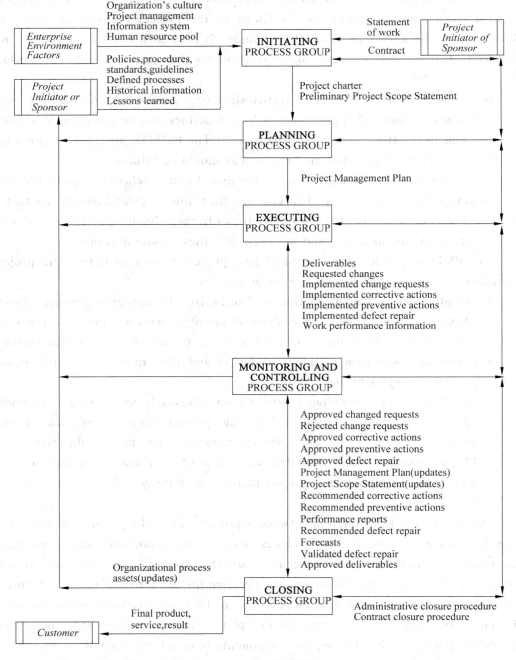

Figure 2.1 The relationship among the PMBOK process groups (taken from Figure 2.3, *PMBOK Guide*, *Fifth Edition*).

The PMBOK identifies nine knowledge areas that a project manager should consider throughout the entire life cycle of a project. Knowledge areas focus on a specific aspect of the overall domain and identify the elements that need to be considered to properly manage a project:

- **Project Integration Management**—This knowledge area looks at the processes and activities needed to identify, define, combine, unify, and coordinate the different actions within a project management process group.
- **Project Scope Management**—This knowledge area handles scope planning, scope definition, creating a WBS (decomposition of the scope into smaller components), scope verification, and scope control.
- **Project Time Management**—This knowledge area concerns five different steps: activity definition, activity sequencing, activity resource estimating, activity duration estimating, and schedule development.
- **Project Cost Management**—This knowledge area involves planning, estimating, budgeting, and controlling costs so a project can be finished within budget.
- **Project Quality Management**—This knowledge area determines policies, objectives, and responsibilities to meet a project's quality standards.
- **Project Human Resource Management**—This knowledge area helps organize and manage a project's team, the people necessary for the completion of the project.
- **Project Communications Management**—This knowledge area involves the processes that ensure timely generation, collection, distribution, storage, retrieval, and disposition of information.
- **Project Risk Management**—This knowledge area envelopes risk management planning, identification, analysis, responses, monitoring, and controlling of a project.
- **Project Procurement Management**—This knowledge area involves the processes necessary to purchase products, services, or results from outside the project team.
- **Project Stakeholder Management**—This knowledge area involves the processes necessary to manage the expectations of and results for the many people and things that can have a stake in the execution and outcome of your project.

The ten knowledge areas are specifically designed to work with the five process groups to identify possible areas for management within the scope of the project. When the two components are combined, they provide guidance for what elements should be considered at what time in a project.

In the context of the Project desktop, the key knowledge areas are scope, time, and cost. These components help you build the initial project schedule framework.

The emphasis for each knowledge area varies by phase of project; some are more important in one phase than another, but all of the nine are used throughout the project.

NOTE

Do not confuse the PMBOK process groups with life cycle phases of projects. This is a common tendency when a project manager tries to decompose a project into logical components. Process groups pertain to all projects; life cycles vary by the type of project, the domain of the work, the complexity and timeframe, and many other factors. Details about phases are covered later in this chapter.

2.2.2 PRINCE2

PRINCE2, which stands for *Projects in Controlled Environments*, is a project management methodology developed by the United Kingdom government. It is in its second release and was originally known as the PRINCE technique. The first release was established in 1989 by the Central Computer and Telecommunications Agency (CCTA) as a standard for information technology project management. Because of its success in IT, the methodology was republished about seven years later in a version that could be applied across many other disciplines. PRINCE2 was again updated in 2005 by the Office of Government Commerce (OGC), has become the standard for project management in the United Kingdom, and is now used in 50 other countries. You can become certified in the use of PRINCE2 at either one of two levels: Foundation and Practitioner.

PRINCE2 uses a simple four-step process to explain what each project needs, as shown in Figure 2.2. This process is explained in more detail using the following eight different processes, sometimes known as the Validation, Quality, Verification, and Approval steps:

- **Start-up**—This is when a project manager is chosen. The need for the project is defined and outlined as to how it will be executed.
- **Direction**—The project manager, who reports to the Project Board, is responsible for managing the details. The Project Board is responsible for the overall success of the current project and defines the direction in which the project will be heading.
- **Initiation**—The Project Initiation Document is prepared and submitted to the Project Board for approval and possible revision.
- **Stage Control**—During this stage, the project is broken down into several different manageable stages. The number of stages depends on the size and risk level of the project, and each stage must also plan for the succeeding stage.

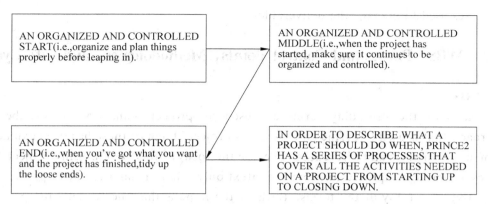

Figure 2.2　The PRINCE2 four-step process.

Before any new stage can begin, the current stage must be fully finished.

- **Stage Boundary Management**—At this stage, the Project Board must review the current stage and then develop the process for the next stage. It is only after the approval for the execution of the current stage and the planning of the next stage that the project can continue.
- **Planning**—This stage is used for deciding what products will be produced and what is required for their production. Then, estimates are made for cost, time, and any other resources, plus any risk analysis, activity scheduling, and process streamlining that is necessary for the project.
- **Product Delivery Management**—This is the production stage, where the project manager confirms that the right goods are being produced correctly and on schedule.
- **Closing**—After everything is finished, the project manager must perform a post-project review, which evaluates the outcome of the project. When this review is approved by the Project Board, the project is complete.

In addition to consideration of these standards and methods, project managers need to understand the environment in which they will be working before they create a schedule. They need to be aware of the various methodologies and approaches that can be used to help them (or confuse them, if they do not understand how and when the methodologies and approaches should be applied). The following section provides an introduction to this information.

NOTE

There is no conflict between PMBOK and PRINCE2. They can be used together if a project manager chooses to use both. PRINCE2 is a methodology and focuses more on deliverables, whereas PMBOK is a standard and focuses on the process and knowledge areas. PRINCE2 establishes a Validation of Process (through a specific focus on deliverables and the activities around them), whereas PMBOK focuses on the processes used to manage the deliverables.

2.3 WBS, Phases and Control Points, Methodologies, and Life Cycles

NOTE

Many of the preceding terms are used by project managers to describe the approach that is used to define and execute a project. Each of these has been explained in many other texts and references. Because the focus of this book is using Project, the topics are brought up here to provide context only; there is no attempt to provide the definitive use of any of the terms. Rather, the hope is that the reader can apply the concepts and techniques as appropriate when building a schedule. The work that needs to happen during a project's life does not automatically conform to a particular methodology.

Before building any schedule, the project manager must consider two key components: work decomposition (what work needs to be done, the Work Breakdown Structure or WBS) and managerial control (stages, phases, and life cycle requirements). The discipline used for either depends on the environment in which the project is executed, so the formality will vary, but both components must be considered. The tasks or activities and milestones (how the work will be accomplished) should not be defined in a project schedule until the WBS and control framework are determined. WBS helps the project manager set parameters around the scope of work to be done; the life cycle sets the controls in place for decisions during project execution. If these two components are kept in control, the project will have a much higher opportunity for success.

2.3.1 Work Breakdown Structure(WBS)

Step one in building a schedule is to begin with a Work Breakdown Structure (WBS) that allows decomposition of the scope of the project from major components to the smallest set of deliverables, called *work packages*. As a best practice, this process is completed before a true schedule is built. It can be done using Project as long as ongoing "use" rules are defined and followed to keep the WBS components intact after the project is approved and baselined.

As mentioned earlier in this chapter, PMI has developed a standard for the WBS. It is a primary component of good project management practices because it forces the discipline of scope definition and control.

If the scope of your project is managed through a WBS, all the tasks and milestones will be created in support of specific work packages and can be rolled up through the structure for tracking progress using Earned Value Management techniques. This practice eliminates some of the common failure points in project management, such as scope creep and fuzzy requirements. All work is clearly linked to the production of a deliverable, and progress against that deliverable can be monitored.

2.3.2 Managerial Control

So many terms are used in the context of managerial control that a few definitions are in order. Hundreds of resources are available to provide detailed explanations; the purpose here is context only. The hope is that these simple descriptions will help the user's understanding when building a project schedule, as discussed in the following sections.

2.3.3 Phases and Gates

Many organizations have established processes for deciding what projects will be approved and for overseeing the projects after they have been launched. In some organizations, the processes are rigorous and robust; in others, the processes might be simple guidelines that have been put in place to help project managers. In either case, a defined set of standard phases and control points (often called *gates*) simplify the decisions that need to be made when running a project. In most cases, templates can be created that standardize the phases and the required control points for different types of projects.

Phases and gates can allow more management control of the project, as they break down the project into smaller components. This helps to keep executive and team focus aligned on the same set of activities. A change between phases is usually defined by some kind of transfer. In many cases, the transfer requires a formal review before the project is allowed to move into the next phase. It is not unusual, however, for phases to begin before the completion of the previous phase, especially when the risks are judged as acceptable. Each organization will make its own determination of the level of control required.

Building the phases and control points into templates is an excellent way to minimize the amount of work that needs to be done when building a new schedule. Many examples are already available in Project, and the organization can build additional ones as needed.

2.3.4 Methodologies

As organizations mature in the project management discipline, they often adopt more formal management control systems. These systems are typically described as methodologies that include processes, rules, standards, and methods for how work will be done. In this section, we identify a few of the ones used in specific industry segments. Each industry has its own set of methodologies, and this chapter does not attempt to identify all of them. The purpose here is to show how managing projects using Project can be included in the methodology to assist in the enforcement and usability of the tools.

2.3.5 Life Cycles

Like methodologies, project life cycles are unique to the industries and disciplines in which they are used. Although all projects have a beginning and an end, they vary greatly in how the work is accomplished. It is nearly impossible to define an ideal life cycle. Some companies and organizations use a single, standardized life cycle for every project, whereas others permit the project manager to choose the best life cycle for the project. In others, a variety of life cycles exists to accommodate different levels of complexity and different styles or types of work.

Regardless of the organization's choices regarding methodologies and life cycles, all organizations can use a scheduling tool to help with project execution. The key to success in every case is that the schedule must be focused on the deliverables to be produced rather than the process. The process must be set up to assist with producing deliverables.

The next section of this chapter provides several examples of methodologies and life cycles in the field of software development to illustrate how Project can be used to enable management of a wide variety of projects.

2.4 Using Microsoft Project with Methodologies and Life Cycles

Almost all organizations have at least a small number of technology projects underway, so software development is an excellent example of the wide variety of project-scheduling approaches available to organizations. The types of projects range from simple to complex, short to multiyear, and goal-oriented to open-ended research.

The following examples discuss the associated software development life cycle (SDLC) and how Project can be set up to support the life cycle. As you review the examples, you should also keep in mind that these projects should be planned and executed using the principles described in the previous sections on project standards (the PMBOK Guide and PRINCE2).

Although strict adherence to the standards is not required or necessary on every project, it is useful to remember that there are five major process groups to be managed on each project and that there are nine knowledge areas that should be considered throughout the project's life cycle.

2.4.1　Waterfall Development Process

Traditional software development is often described as a waterfall model because it is sequential in nature. The assumption with this model is that phases can be completed in order with little or no need to repeat the previous activities. Development is described as a waterfall, steadily falling down through traditional phases such as definition, preliminary design, detailed design, coding, testing, implementation, and transition to operations.

This method of development is used in many organizations today, especially those involved in multiyear programs. The phases can be lengthy and the work can be exacting. Although the name suggests that all work from one phase is completed before moving into the next phase, these types of projects are often set up with overlapping phases so that design can begin on certain deliverables as soon as the definition of the work for those deliverables is completed. In addition, there is typically some level of iterative development involved in almost all projects, but the term "waterfall" is still in common use today.

In this type of project, the tendency is to set up the project schedule in the same order as the major phase names. Instead, the project can be set up so that it is broken into logical work packages that can be monitored and measured separately.

2.4.2　Iterative Development

Iterative development provides a strong framework for planning purposes and also flexibility for successive iterations of software development. The Rational Unified Process (RUP) and the Dynamic Systems Development Methods are two frameworks for this type of project life cycle. RUP is not only a methodology for software engineering project management; it also has a set of software tools for using the specific methodology that is the Rational Unified Process. Figure 2.3 depicts the RUP workflow.

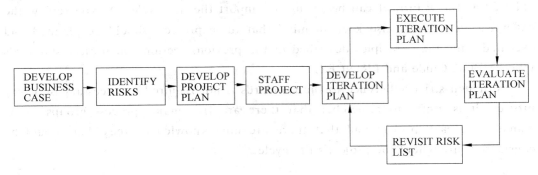

Figure 2.3　RUP workflow.

　　The goal for this type of software development life cycle (SDLC) is to allow the developers to learn through incremental development and the use of prototypes instead of trying to complete detailed requirements before the development work begins. Agile and XP can also be considered to be iterative methods.

1. Agile Development Process

　　Agile is a philosophy of project management that moves away from the classic project management methods and focuses less on planning and more on execution. In Agile, crucial decisions are made during the project execution phase, instead of the planning phase. As business and project environments become more fluid and dynamic, the amount of time for effective planning becomes less and less. This does not mean that planning is ignored; rather, the focus shifts to supporting decisions during project execution instead of finalizing all decisions during the planning stage.

　　Agile is not an all-or-nothing methodology either; it is possible to combine Agile with more classic project management ideas. Whereas classic project management is comprehensive and works in diverse situations, Agile can add various ideas for facing new and unique situations that can be found in creative, knowledge-based industries.

　　Here are some of the attributes of an Agile SDLC:

- Short development cycles are used to produce working software in weeks rather than months.
- Communication between the business users and the developers occurs daily.
- Documentation of working functionality is captured after the software is completed; there is limited documentation of the requirements or design.
- Timeboxing is used to force tough decisions early in the project.
- Changes to requirements are expected; they are a result of early working prototypes and are a goal of the process.
- The project manager for an Agile team is focused on ensuring excellent

communication as the primary mechanism to maintain progress.

Agile development can be difficult for large organizations to embrace because it does not require a focus on formal planning of an entire project.

The major difference is that the primary measurement of progress is frequent delivery of small amounts of working software. With a focus on feature delivery, it can sometimes be difficult to understand the overall picture, so strong project management must provide this clarity.

CAUTION

The use of Agile should not be used as an excuse to avoid planning or managing a budget. The approach is meant to provide a lighter and faster method to reach a goal, but the goal is still required.

In this type of environment, a project team can still use Project to support its goals. In an Agile environment, the tool is not used to develop a robust schedule with a beginning-to-end flow of tasks and resources. Its use in this case supports communication to management and ensures that changes are captured and the backlog of work is moved through each successive iteration of the project schedule.

In the following example, the project manager has established a budget summary task to provide rollup of budget for management purposes. Successive sets of work are defined in small iterations, while the overall timeframe and budget are obvious for all (see Figure 2.4). This approach enables the team to perform iterative planning while still meeting the business requirements of not exceeding a specific timeframe and budget.

By establishing a project schedule with an overall goal, the needs of the team and their management can be met. Refer to Figure 2.4 for an example of a short project that is expected to complete within a target effort of 340 hours. The work is not fully defined at the beginning so that the team has the flexibility to decide what work will happen in what order. Management is still able to see overall metrics of planned work, actual work, and the current estimate of work remaining.

Agile is an extremely successful method of software development that is well suited to an environment with self-motivated teams, open communications, and leadership that is comfortable with a prototyping approach to work. It does not fit all projects, but when it works, it works well.

The schedule created in Project for this type of approach becomes a tool for communication, overall budget and time goals, and historical tracking purposes.

2. Extreme Programming

Extreme Programming, or XP, is another method within the Agile family that has become a simple and flexible way for developing software through the writing of

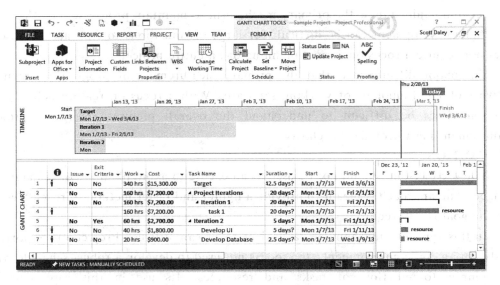

Figure 2.4　An Agile project showing overall budget, work, and timeframe with iterative development.

tests. It is designed to be used by a group of two to ten programmers who are able to execute tests in a fraction of a day. It uses short cycles of phases, with early and continuing feedback during each cycle. This flexibility enables it to respond to changing business demands through the implementation of functionality.

NOTE

For organizations that use the Project Server, this method enables them to use an Agile approach and yet have oversight of the entire project portfolio. Agile projects coexist with standard iterative projects in their Project Server environment; the projects have planned timeframes, resources, and budgets but are not required to have all the work scoped out at the beginning of the project.

XP's use of automated tests, written by the programmers to scrutinize development, helps in early detection of defects and also enables the cycle of phases to evolve as the project continues. These automated tests depend both on the short-term instincts of programmers and also on the long-term interests of the project. XP also relies heavily on a system of oral communication, tests, and source code to help communicate the system structure and intent.

These processes allow for the day-to-day programming of a feature, and then moving on to testing, implementation, design, and integration, all packed into each cycle. The scheduling methods used in the preceding Agile example can again be adapted for XP.

3. Spiral Development Project

Spiral development was defined by Barry Boehm in 1985 and is often used in fairly large projects that take months to two years or more to complete. The initial focus might be on core functionality, and then the "bells and whistles" such as graphical user interfaces and reporting are added at a later time. This is sometimes considered to be another form of iterative development, but the structure of the plans and schedule focus on a robust core design in the early stages.

2.4.3 Research Project

A research project might be the most difficult type of project to tackle when it comes to constructing a project schedule. Often there is no clear goal in mind, and there might not even be an expectation of a specific end date or budget. On the other side, however, even research projects must be funded by someone, and they must have a working staff, so there is typically some expectation of a result. In most cases, there is also an expectation that the funding is used responsibly, so there must be a process in place to track how the money has been spent.

Project can once again be used to support this type of project as a tracking mechanism and a place to bring together the set of work that will be performed. The schedule will not require all the advanced features of critical path analysis, resource leveling, and predecessor/successor relationships, but it can be used as an easy method of historical support and a loose prediction of the work that is to be accomplished.

2.5 Accommodating Teaming Styles

High-performance teams, self-managed teams, and other nontraditional structures began to emerge more than 50 years ago in Great Britain and gained acceptance across the globe as several large corporations began to adopt the concepts. The general idea behind these teaming styles was to loosen managerial constraints in an effort to increase worker performance and make quantum leaps in accomplishment of organizational goals.

When framed correctly, the teams need little direction and excel in accomplishing the goals of their projects. If the dynamics are not understood, however, little is accomplished. From a project management perspective, Agile or XP projects can be a bit intimidating because the team dynamics can overwhelm the designated leader. In reality, successful self-managed teams are not leaderless. They have simply figured out a mechanism to allow many people within the team to play a leadership role.

Even in a team where a project management role has not been defined, someone must take on the job of setting a direction to accomplish a goal. The goal might only be one week in the future, but the team must coalesce around that goal, and the person who makes that happen is a leader. If the project manager understands the dynamics of the team, he or she can use these dynamics to improve the team's focus and increase its performance. The PM must be comfortable with sharing decision making and needs to focus heavily on communication of information within the team and with the stakeholders of the project. Things change quickly in this environment, so communication of status becomes a driving force for the project.

Project is an excellent tool to aid the PM in communication. Two components need to be established to make this successful. The overall goal of the project needs to be clear to the team, and the boundaries of the project (overall timeframe, scope, resources, and budget) must be understood. If these components are established within the tool as a baseline, the remainder of the schedule can be flexible or rigid, as dictated by the project structure and the teaming style.

2.6 Consultants' Tips

2.6.1 Determine the Approach to Use in Managing Your Project

Project has a rich set of features that enable the project manager and team to track projects at a detailed level. It also has enough flexibility to allow high-level tracking without a demand for the detail. It can be and has been used to support all industries and all domains within those industries.

Because the software has so many capabilities, it must be well understood to be used correctly. The scheduling engine anticipates your needs and moves the dates or adjusts the amount of work that is to be accomplished based on the parameters that you set. Because it does this, project managers must have a clear understanding of the approach that they want to use on their projects before they begin entering tasks.

2.6.2 Use WBS as a First Step in Project Definition

Always start with a WBS to help you be clear on the goals of the project. Wait to add the task-level details until you are sure that you have decomposed the WBS to the work package level that is right for the type of project you are leading. Do not confuse the listing of activities with the completion of deliverables.

2.6.3 Use the 5×9 Checklist for Planning

Remember the 5×9 checklist and consider it when planning and executing each project. As you move through the phases of your project's life cycle, spend a moment to consider which of the five process groups is most dominant at the moment and which of the nine knowledge areas plays the most important part in the project's evolution.

2.7 New Words and Expressions

milestone	n. 里程碑,划时代的事件
infrastructure	n. 基础设施;公共建设
under budget	在预算之内
DARPA(Defense Advanced Research Projects Agency)	abbr. 美国国防部高级研究计划局
NASA(National Aeronautics and Space Administration)	abbr. 美国国家航空和宇宙航行局
CPM(critical path method)	abbr. 关键路径法
PERT(Program Evaluation and Review Technique)	abbr. 统筹法;程序估计和检查技术
methodologies	n. 方法论
PMBOK(Project Management Body of Knowledge)	abbr. 项目管理知识体系
software development life cycle	软件开发生命周期
waterfall model	n. 瀑布模型
rational unified process	统一软件开发过程
flexibility	n. 灵活性;适应性
prototyping approach	n. 原型法
project portfolio	n. 项目组合
scrutinize	vt. 详细检查;细看
oral communication	口头沟通;口语交谈
bells and whistles	附加的修饰物;花里胡哨
iterative development	迭代开发
a bit intimidating	让人有点惶恐
overwhelm	vt. 压倒;压垮
coalesce	vt. 使…联合;使…合并
flexibility	n. 灵活性;适应性
evolution	n. 演变;进展

2.8　Questions

Single Choice Questions

1. What does the acronym DARPA stand for? (　　)

    A. Defense Advanced Research Projects Agent

    B. Defense Advantageous Research Projects Agent

    C. Defense Advantageous Research Projects Agency

    D. Defense Advanced Research Projects Agency

2. Two other major developments for project management to grow out of this period were(　　).

    A. the Program Evaluation and the Review Technique

    B. the Critical Path Method and the Review Technique

    C. the Critical Path Method and the Program Evaluation and Review Technique

    D. the Critical Path Method and the Program Evaluation Technique

3. Defines and refines objectives, and plans the course of action required to attain the objectives and scope that the project is to address. What does that sentence describe about? (　　)

    A. Initiating Process Group

    B. Planning Process Group

    C. Executing Process Group

    D. Monitoring and Controlling Process Group

4. Regularly measures and monitors progress to identify variances from the project management plan so that corrective action can be taken when necessary to meet project objectives. What does that sentence describe about? (　　)

    A. Initiating Process Group

    B. Planning Process Group

    C. Executing Process Group

    D. Monitoring and Controlling Process Group

5. The (　　) is one component of the plan that focuses on the timeline for the activities to be performed.

    A. plan　　　　B. schedule　　　　C. initiate　　　　D. monitor

6. This knowledge area looks at the processes and activities needed to identify, define, combine, unify, and coordinate the different actions within a project management process group. What does that sentence describe about? (　　)

    A. Project Cost Management

    B. Project Time Management

C. Project Integration Management

D. Project Scope Management

7. This is when a project manager is chosen. The need for the project is defined and outlined as to how it will be executed. What does that sentence describe about? ( )

    A. Initiation   B. Start-up   C. Closing   D. Planning

8. These systems are typically described as ( ) that include processes, rules, standards, and methods for how work will be done.

    A. methodologies   B. work breakdown structure

    C. phases and gates   D. life cycles

9. The follow is not the attributes of an Agile SDLC? ( )

    A. Short development cycles are used to produce working software in weeks rather than months

    B. Communication between the business users and the developers occurs daily

    C. Documentation of working functionality is captured after the software is completed; there is limited documentation of the requirements or design

    D. Agile is a philosophy of project management that moves away from the classic project management methods and focuses more on planning and more on execution

10. Consultants' Tips does not include ( ).

    A. Using Microsoft Project with Methodologies and Life Cycles

    B. Determine the Approach to Use in Managing Your Project

    C. Use the 5×9 Checklist for Planning

    D. Use WBS as a First Step in Project Definition

2.9 Problems

After reading this chapter and completing the exercises, you will be able to do the following:

1. Describe project management and project management industry standards.
2. Discuss project management body of knowledge.
3. Explain WBS, Phases and Control Points, Methodologies, and Life Cycles
4. How using Microsoft project with methodologies and life cycles.
5. Explain accommodating teaming styles.
6. Discuss the determination of the approach to use in managing your project.

Unit 3　Database—NoSQL Databases: An Overview

Pramod Sadalage provides an overview of NoSQL databases, explaining what NoSQL is, types of NoSQL databases, and why and how to choose a NoSQL database.

Over the last few years we have seen the rise of a new type of databases, known as NoSQL databases, which are challenging the dominance of relational databases. Relational databases have dominated the software industry for a long time providing mechanisms to store data persistently, concurrency control, transactions, mostly standard interfaces and mechanisms to integrate application data, reporting. The dominance of relational databases, however, is cracking.

3.1　NoSQL: What does it mean

What does NoSQL mean and how do you categorize these databases? NoSQL means Not Only SQL, implying that when designing a software solution or product, there are more than one storage mechanism that could be used based on the needs. NoSQL was a hashtag (#nosql) choosen for a meetup to discuss these new databases. The most important result of the rise of NoSQL is Polyglot Persistence. NoSQL does not have a prescriptive definition but we can make a set of common observations, such as:
- Not using the relational model
- Running well on clusters
- Mostly open-source
- Built for the 21st century web estates
- Schema-less

3.2　Why NoSQL Databases

An example is shown in Figure 3.1, application developers have been frustrated with the impedance mismatch between the relational data structures and the in-memory data structures of the application. Using NoSQL databases allows developers to develop without having to convert in-memory structures to relational structures.

There is also movement away from using databases as integration points in favor of encapsulating databases with applications and integrating using services as shown in Figure 3.2.

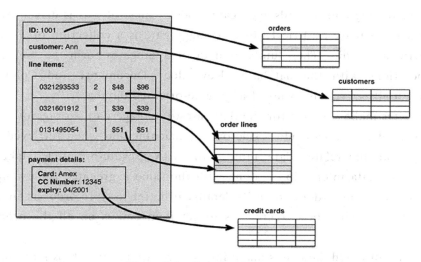

Figure 3.1 Relational structures

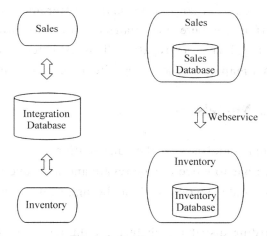

Figure 3.2 Data storage

The rise of the web as a platform also created a vital factor change in data storage as the need to support large volumes of data by running on clusters.

Relational databases were not designed to run efficiently on clusters.

The data storage needs of an ERP application are lot more different than the data storage needs of a Facebook or an Etsy, for example.

3.3 Aggregate Data Models

Relational database modelling is vastly different than the types of data structures that application developers use. Using the data structures as modelled by the developers to solve different problem domains has given rise to movement away from relational modelling and towards aggregate models, most of this is driven by Domain Driven Design, a book by Eric Evans. An aggregate is a collection of data that we interact with as a unit. These units of data or aggregates form the boundaries for ACID operations with the database, Key-value, Document, and Column-family databases can all be seen as forms of aggregate-oriented database.

Aggregates make it easier for the database to manage data storage over clusters, since the unit of data now could reside on any machine and when retrieved from the database gets all the related data along with it. Aggregate-oriented databases work best when most data interaction is done with the same aggregate, for example when there is need to get an order and all its details, it better to store order as an aggregate object but dealing with these aggregates to get item details on all the orders is not elegant.

Aggregate-oriented databases make inter-aggregate relationships more difficult to handle than intra-aggregate relationships. Aggregate-ignorant databases are better when interactions use data organized in many different formations. Aggregate-oriented databases often compute materialized views to provide data organized differently from their primary aggregates. This is often done with map-reduce computations, such as a map-reduce job to get items sold per day.

3.4 Distribution Models

Aggregate oriented databases make distribution of data easier, since the distribution mechanism has to move the aggregate and not have to worry about related data, as all the related data is contained in the aggregate. There are two styles of distributing data:

- Sharding: Sharding distributes different data across multiple servers, so each server acts as the single source for a subset of data.
- Replication: Replication copies data across multiple servers, so each bit of data can be found in multiple places. Replication comes in two forms,
    (1) Master-slave replication makes one node the authoritative copy that handles writes while slaves synchronize with the master and may handle reads.
    (2) Peer-to-peer replication allows writes to any node; the nodes coordinate to synchronize their copies of the data.

Master-slave replication reduces the chance of update conflicts but peer-to-peer replication avoids loading all writes onto a single server creating a single point of failure. A system may use either or both techniques. Like Riak database shards the data and also replicates it based on the replication factor.

3.5 CAP theorem

In a distributed system, managing consistency(C), availability(A) and partition toleration(P) is important. Eric Brewer put forth the CAP theorem which states that in any distributed system we can choose only two of consistency, availability or partition tolerance. Many NoSQL databases try to provide options where the developer has choices where they can tune the database as per their needs. For example if you consider Riak a distributed key-value database. There are essentially three variables r, w, n where

- r = number of nodes that should respond to a read request before its considered successful.
- w = number of nodes that should respond to a write request before its considered successful.
- n = number of nodes where the data is replicated aka replication factor.

In a Riak cluster with 5 nodes, we can tweak the r,w,n values to make the system very consistent by setting r = 5 and w = 5 but now we have made the cluster susceptible to network partitions since any write will not be considered successful when any node is not responding. We can make the same cluster highly available for writes or reads by setting r = 1 and w = 1 but now consistency can be compromised since some nodes may not have the latest copy of the data. The CAP theorem states that if you get a network partition, you have to trade off availability of data versus consistency of data. Durability can also be traded off against latency, particularly if you want to survive failures with replicated data.

NoSQL databases provide developers lot of options to choose from and fine tune the system to their specific requirements. Understanding the requirements of how the data is going to be consumed by the system, questions such as is it read heavy vs write heavy, is there a need to query data with random query parameters, will the system be able handle inconsistent data.

Understanding these requirements becomes much more important, for long we have been used to the default of RDBMS which comes with a standard set of features no matter which product is chosen and there is no possibility of choosing some features over other. The availability of choice in NoSQL databases, is both good and bad at the same time. Good because now we have choice to design the system according to the requirements. Bad because now you have a choice and we have to make a good choice based on requirements and there is a chance where the same database product may be used properly or not used properly.

An example of feature provided by default in RDBMS is transactions, our development methods are so used to this feature that we have stopped thinking about what would happen when the database does not provide transactions. Most NoSQL databases do not provide transaction support by default, which means the developers have to think how to implement transactions, does every write have to have the safety of transactions or can the write be segregated into "critical that they succeed" and "its okay if I lose this write" categories. Sometimes deploying external transaction managers like ZooKeeper can also be a possibility.

3.6 Types of NoSQL Databases

3.6.1 Key-Value databases

Key-value stores are the simplest NoSQL data stores to use from an API perspective. The client can either get the value for the key, put a value for a key, or delete a key from the data store. In the case shown in Figure 3.3, The value is a blob that the data store just stores, without caring or knowing what's inside; it's the responsibility of the application to understand what was stored. Since key-value stores always use primary-key access, they generally have great performance and can be easily scaled.

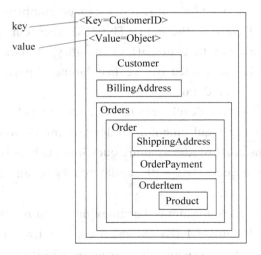

Figure 3.3   Key-Value databases

Some of the popular key-value databases are Riak, Redis (often referred to as Data Structure server), Memcached and its flavors, Berkeley DB, HamsterDB (especially suited for embedded use), Amazon DynamoDB (not open-source), Project Voldemort and Couchbase.

All key-value databases are not the same, there are major differences between these products, for example: Memcached data is not persistent while in Riak it is, these features are important when implementing certain solutions. Let's consider we need to implement caching of user preferences, implementing them in memcached means when the node goes down all the data is lost and needs to be refreshed from source system, if we store the same data in Riak we may not need to worry about losing data but we must also consider how to update stale data. It's important to not only choose a key-value database based on your requirements, it's also important to choose which key-value database.

3.6.2 Document databases

Documents are the main concept in document databases, as shown in Figure 3.4. The database stores and retrieves documents, which can be XML, JSON, BSON, and so on. These documents are self-describing, hierarchical tree data structures which can consist of maps, collections, and scalar values. The documents stored are similar to each other but do not have to be exactly the same. Document databases store documents in the value part of the key-value store; think about document databases as key-value stores where the value is examinable. Document databases such as MongoDB provide a rich query language and constructs such as database, indexes etc allowing for easier transition from relational databases.

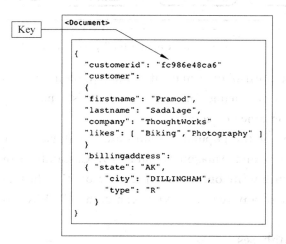

Figure 3.4   Document databases

Some of the popular document databases we have seen are MongoDB, CouchDB, Terrastore, OrientDB, RavenDB, and of course the well-known and often reviled Lotus Notes that uses document storage.

3.6.3 Column family stores

Column-family databases store data in column families as rows that have many columns associated with a row key. Column families are groups of related data that is often accessed together. For a Customer, we would often access their Profile information at the same time, but not their Orders.

An example is shown in Figure 3.5, each column family can be compared to a container of rows in an RDBMS table where the key identifies the row and the row consists of multiple columns. The difference is that various rows do not have to have the same columns, and columns can be added to any row at any time without having to add it to other rows.

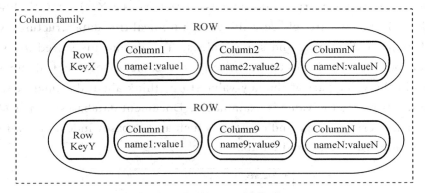

Figure 3.5 Column-family databases

When a column consists of a map of columns, then we have a super column. A super column consists of a name and a value which is a map of columns. Think of a super column as a container of columns.

Cassandra is one of the popular column-family databases; there are others, such as HBase, Hypertable, and Amazon DynamoDB. Cassandra can be described as fast and easily scalable with write operations spread across the cluster. The cluster does not have a master node, so any read and write can be handled by any node in the cluster.

3.6.4 Graph Databases

Graph databases allow you to store entities and relationships between these entities. Entities are also known as nodes, which have properties. Think of a node as an instance of an object in the application, as shown in Figure 3.6. Relations are known as edges that can have properties. Edges have directional significance; nodes are organized by relationships which allow you to find interesting patterns between the nodes. The organization of the graph lets the data to be stored once and then interpreted in different ways based on relationships.

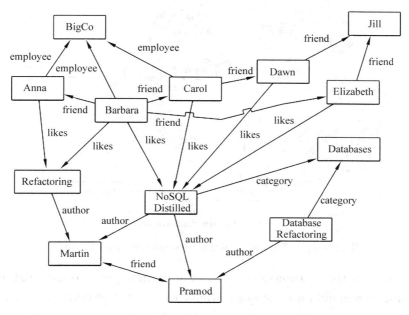

Figure 3.6　Graph databases

　　Usually, when we store a graph-like structure in RDBMS, it's for a single type of relationship ("who is my manager" is a common example). Adding another relationship to the mix usually means a lot of schema changes and data movement, which is not the case when we are using graph databases. Similarly, in relational databases we model the graph beforehand based on the Traversal we want; if the Traversal changes, the data will have to change.

　　In graph databases, traversing the joins or relationships is very fast. The relationship between nodes is not calculated at query time but is actually persisted as a relationship. Traversing persisted relationships is faster than calculating them for every query.

　　In the case shown in Figure 3.7, nodes can have different types of relationships between them, allowing you to both represent relationships between the domain entities and to have secondary relationships for things like category, path, time-trees, quad-trees for spatial indexing, or linked lists for sorted access. Since there is no limit to the number and kind of relationships a node can have, they all can be represented in the same graph database.

　　Relationships are first-class citizens in graph databases; most of the value of graph databases is derived from the relationships. Relationships don't only have a type, a start node, and an end node, but can have properties of their own. Using

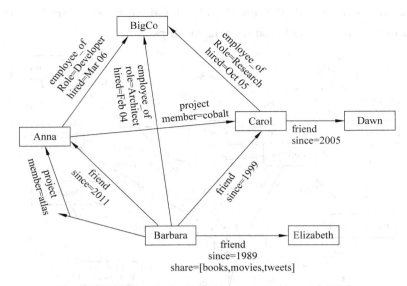

Figure 3.7 The relationships between the domain entities

these properties on the relationships, we can add intelligence to the relationship—for example, since when did they become friends, what is the distance between the nodes, or what aspects are shared between the nodes. These properties on the relationships can be used to query the graph.

Since most of the power from the graph databases comes from the relationships and their properties, a lot of thought and design work is needed to model the relationships in the domain that we are trying to work with. Adding new relationship types is easy; changing existing nodes and their relationships is similar to data migration, because these changes will have to be done on each node and each relationship in the existing data.

There are many graph databases available, such as Neo4J, Infinite Graph, OrientDB, or FlockDB (which is a special case: a graph database that only supports single-depth relationships or adjacency lists, where you cannot traverse more than one level deep for relationships).

3.7 Why choose NoSQL database

We've covered a lot of the general issues you need to be aware of to make decisions in the new world of NoSQL databases. It's now time to talk about why you would choose NoSQL databases for future development work. Here are some broad reasons to consider the use of NoSQL databases.

- To improve programmer productivity by using a database that better matches an application's needs.

- To improve data access performance via some combination of handling larger data volumes, reducing latency, and improving throughput.

It's essential to test your expectations about programmer productivity and/or performance before committing to using a NoSQL technology. Since most of the NoSQL databases are open source, testing them is a simple matter of downloading these products and setting up a test environment.

Even if NoSQL cannot be used as of now, designing the system using service encapsulation supports changing data storage technologies as needs and technology evolve. Separating parts of applications into services also allows you to introduce NoSQL into an existing application.

3.8 Choosing NoSQL database

Given so much choice, how do we choose which NoSQL database? As described much depends on the system requirements, here are some general guidelines:
- Key-value databases are generally useful for storing session information, user profiles, preferences, shopping cart data. We would avoid using Key-value databases when we need to query by data, have relationships between the data being stored or we need to operate on multiple keys at the same time.
- Document databases are generally useful for content management systems, blogging platforms, web analytics, real-time analytics, and ecommerce-applications. We would avoid using document databases for systems that need complex transactions spanning multiple operations or queries against varying aggregate structures.
- Column family databases are generally useful for content management systems, blogging platforms, maintaining counters, expiring usage, heavy write volume such as log aggregation. We would avoid using column family databases for systems that are in early development, changing query patterns.
- Graph databases are very well suited to problem spaces where we have connected data, such as social networks, spatial data, routing information for goods and money, recommendation engines.

3.9 Schema-less ramifications

All NoSQL databases claim to be schema-less, which means there is no schema enforced by the database themselves. Databases with strong schemas, such as relational databases, can be migrated by saving each schema change, plus its data migration, in a version-controlled sequence. Schema-less databases still need careful migration due to the implicit schema in any code that accesses the data.

Schema-less databases can use the same migration techniques as databases with strong schemas, in schema-less databases we can also read data in a way that's tolerant to changes in the data's implicit schema and use incremental migration to update data, thus allowing for zero downtime deployments, making them more popular with 24 * 7 systems.

3.10 Conclusion

All the choice provided by the rise of NoSQL databases does not mean the demise of RDBMS databases. We are entering an era of polyglot persistence, a technique that uses different data storage technologies to handle varying data storage needs. Polyglot persistence can apply across an enterprise or within a single application.

For more details, read NoSQL Distilled: A Brief Guide to the Emerging World of Polyglot Persistence by Pramod Sadalage and Martin Fowler. A version of this article was originally published on ThoughtWorks' Big Data Analytics blog.

3.11 New Words and Expressions

concurrency control	并发控制
categorize	vt. 分类
clusters	n. 群集
ERP(Enterprise Resource Planning)	abbr. 企业资源计划
aggregate	adj. 聚合的;集合的
materialized view	实体化视图
sharding	n. 分片;分区
RDBMS(Relational Database Management System)	abbr. 关系型数据库管理系统
retrieve	vt. 检索;恢复
hierarchical	adj. 分层的;等级体系的
Spatial indexing	空间索引
be derived from	来源于
reduce latency	减少延迟
encapsulation	n. 封装;包装
Schema-less ramifications	n. 无模式的影响

3.12 Questions

Single Choice Questions

1. NoSQL does not have a prescriptive definition but we can make a set of

common observations, such as:(　　).

　　A. Mostly open-source

　　B. Running well on clusters

　　C. Using the relational model

　　D. Built for the 21st century web estates

2. The most important result of the rise of NoSQL is(　　).

　　A. Tokyo Cabinet　　　　　　B. Polyglot Programming

　　C. Voldemort　　　　　　　　D. Polyglot Persistence

3. Which of the following databases can not be seen as forms of aggregate-oriented database? (　　)

　　A. Key-value databases　　　　B. Graph Databases

　　C. Document databases　　　　D. Column-family databases

4. Which of the following expression is not true? (　　)

　　A. Aggregate-oriented databases do not work well when most data interaction is done with the same aggregate

　　B. Aggregate-oriented databases make inter-aggregate relationships more difficult to handle than intra-aggregate relationships

　　C. Aggregate-ignorant databases are better when interactions use data organized in many different formations

　　D. Aggregate-oriented databases often compute materialized views to provide data organized differently from their primary aggregates

5. What are the two styles of distributing data? (　　)

　　A. Sharding and Master-slave Replication

　　B. Sharding and Replication

　　C. Peer-to-peer Replication and Master-slave Replication

　　D. Sharding and Peer-to-peer Replication

6. What does the acronym CAP stand for? (　　)

　　A. consistency, availability and partition

　　B. consistency, available and partition toleration

　　C. consistence, availability and partition toleration

　　D. consistency, availability and partition toleration

7. Which sentence is true according to the article? (　　)

　　A. Durability can also be traded off against latency, particularly if you want to survive failures with replicated data

　　B. Consistency can also be traded off against latency, particularly if you want to survive failures with replicated data

　　C. Availability can also be traded off against latency, particularly if you want to survive failures with replicated data

D. Partition toleration can also be traded off against latency, particularly if you want to survive failures with replicated data

8. As described much depends on the system requirements, which of the following general guidelines is not true? (　　)

    A. Key-value databases are generally useful for storing session information, user profiles, preferences, shopping cart data. We would avoid using Key-value databases when we need to query by data, have relationships between the data being stored or we need to operate on multiple keys at the same time

    B. Document databases are generally useful for content management systems, blogging platforms, web analytics, real-time analytics, and ecommerce-applications. We would avoid using document databases for systems that need complex transactions spanning multiple operations or queries against varying aggregate structures

    C. Relational databases are generally useful for content management systems, blogging platforms, maintaining counters, expiring usage, heavy write volume such as log aggregation

    D. Graph databases are very well suited to problem spaces where we have connected data, such as social networks, spatial data, routing information for goods and money, recommendation engines

9. Which of the following sentence is true? (　　)

    A. All NoSQL databases claim to be schema-less, which means there is no schema enforced by the database themselves

    B. All NoSQL databases claim to be schema-more, which means there is no schema enforced by the database themselves

    C. All NoSQL databases claim to be schema-less, which means there is schema enforced by the database themselves

    D. All NoSQL databases claim to be schema-more, which means there is schema enforced by the database themselves

10. Which of the following sentence is true? (　　)

    A. Databases with strong schemas, such as non-relational databases, cannot be migrated by saving each schema change, plus its data migration, in a version-controlled sequence

    B. Databases with strong schemas, such as non-relational databases, can be migrated by saving each schema change, plus its data migration, in a version-controlled sequence

    C. Databases with strong schemas, such as relational databases, can be migrated by saving each schema change, plus its data migration, in a version-controlled sequence

D. Databases with strong schemas, such as relational databases, cannot be migrated by saving each schema change, plus its data migration, in a version-controlled sequence

3.13 Problems

After reading this chapter and completing the exercises, you will be able to do the following:

1. What does NoSQL mean?
2. Explain Aggregate data model and distribution model.
3. Explain CAP theorem.
4. Distinguish between all types of NoSQL databases.
5. Why choose NoSQL database?
6. Discuss schema-less ramifications.

Unit 4   Programming—What Software Architects Need to Know About DevOps

Much has been written about DevOps, but most of it focuses on the Ops side of things. This article highlights the most important aspects for software architects and engineers. As such, the authors will cover core aspects that software architects should be aware of: DevOps, its motivation and its main practices, organizational aspects of introducing DevOps, and implications for software architecture.

4.1   What Software Architects Need to Know About DevOps

The motivation for DevOps is that developers and operators often have opposing goals: Developers (Devs) try to push new features into the product, while the core concern of operators (Ops) is system dependability and availability—which is facilitated by putting high burdens on the process of releasing new features. *The goal of DevOps is to achieve both, high frequencies of releasing new features and dependability*, by encouraging collaboration and shared responsibilities between Devs and Ops.

If your organization is interested in achieving shorter cycles for new features becoming available in software products, without sacrificing quality, then you should care about DevOps. Take the example of IBM, which, by applying DevOps and related practices, "… has gone from spending about 58% of its development resources on innovation to about 80%," according to this recent report. Wouldn't you like to spend most of your time on the fun and innovative parts of your job?

DevOps brings about changes on several levels: organizational structure and culture, software structure, test automation, and continuous deployment. Architects need to know about DevOps implication on team structure, stakeholders, and architecture styles and patterns. We start with a definition of DevOps, before we give an overview of each of those topics.

4.2   Defining DevOps

In our book, we use a goal-driven definition of DevOps:
- DevOps is a set of practices intended to reduce the time between committing a change to a system and the change being placed into normal production, while

ensuring high quality.

This definition has several implications. First, the *quality of the deployed change* to a system (usually in the form of code) is important. *Quality* means suitability for use by various stakeholders including end users, developers, or system administrators. It also includes availability, security, reliability, and other "ilities." One method for ensuring quality is to have a variety of automated test cases that must be passed prior to placing changed code into production. Another method is to test the change in production with a limited set of users prior to opening it up to the world. Still another method is to closely monitor newly deployed code for a period of time. We do not specify in the definition how quality is ensured, but we do require that production code be of high quality.

The definition also requires the *delivery mechanism to be of high quality*. This implies that reliability and the repeatability of the delivery mechanism should be high. If it fails regularly, the time required increases. If there are errors in how the change is delivered, the quality of the deployed system suffers, e.g., through reduced availability or reliability.

We identify *two times* as being important. One is the time when a developer commits newly developed code. This marks the end of basic development and the beginning of the deployment path. The second time is the deploying of that code into production. There is a period after code has been deployed into production when the code undergoes live testing and is closely monitored for potential problems. Once the code has passed live testing and close monitoring, then it is considered as a portion of the normal production system. We make a distinction between deploying code into production for live testing and close monitoring and then, after passing the tests, promoting the newly developed code to be equivalent to previously developed code.

Our definition is *goal-oriented*. We do not specify the form of the practices or whether tools are used to implement them. If a practice is intended to reduce the time between a commit from a developer and deploying into production, it is a DevOps practice—whether it involves agile methods, tools, or forms of coordination. This is in contrast to several other definitions. Wikipedia, for example, stresses communication, collaboration, and integration between various stakeholders without stating the goal of such communication, collaboration, or integration. Timing goals are implicit. Other definitions stress the connection between DevOps and agile methods. Again, there is no mention of the purpose of utilizing agile methods on either the time to develop or the quality of the production system. Still other definitions stress the tools being used, without mentioning the goal of DevOps practices, time involved, or quality.

Finally, the goals specified in the definition do not restrict the *scope of DevOps* practices to testing and deployment. In order to achieve these goals, it is important to include an Ops perspective in the collection of requirements, i.e., significantly earlier than committing changes. Analogously, the definition does not mean DevOps practices end with deployment into production; the goal is to ensure high quality of the deployed system throughout its lifecycle. Thus, monitoring practices that help achieve the goals are to be included as well.

4.3 DevOps Practices and Architectural Implications

We have identified five different categories of DevOps practices, which satisfy our definition. For each practice, we discuss the architectural implications.

- **Treat Ops as first-class citizens** from the point of view of requirements. These practices fit in the high quality aspect of the definition. Operations have a set of requirements that pertain to logging and monitoring. For example, logging messages should be understandable and useable by an operator. Involving operations in the development of requirements will ensure that these types of requirements are considered.

  Adding requirements to a system from Ops may require some architectural modification. In particular, the Ops requirements are likely to be in the area of logging, monitoring, and information to support incident handling. These requirements will be like other requirements for modifications to a system: possibly requiring some minor modifications to the architecture but, typically, not drastic modifications.

- **Make Dev more responsible for relevant incident handling.** These practices are intended to shorten the time between the observation of an error and the repair of that error. Organizations that utilize these practices typically have a period of time in which Dev has primary responsibility for a new deployment; later on Ops has primary responsibility.

  By itself, this change is just a process change and should require no architectural modifications. However, just as with the previous practice, once Dev becomes aware of the requirements for incident handling, some architectural modifications may result.

- **Continuous deployment.** Practices associated with continuous deployment are intended to shorten the time between a developer committing code to a repository and that code being deployed. Continuous deployment also emphasizes automated tests, to increase the quality of code making its way into production. Continuous deployment is the practice which leads to the most far-reaching architectural modifications. On the one hand, an organization can introduce continuous deployment practices with no major architectural changes. On the other hand, organizations that have adopted continuous deployment practices frequently begin moving to a microservice architecture. We cover microservice architectures and explore the reasons for adoption below.

- **Develop infrastructure code with the same set of practices as application code.** Practices that apply to the development of infrastructure code are intended both to ensure high quality in the deployed applications and to ensure that deployments proceed as planned. Errors in deployment scripts such as misconfigurations can cause errors in the application, in the environment, or in the deployment process. Applying quality control practices used in normal software development when developing operations scripts and processes will help control the quality of these specifications. These practices will not affect the application code but may affect the architecture of the infrastructure code.

- **Enforced deployment process used by all,** including Dev and Ops personnel. These practices are intended to ensure higher quality of deployments, e.g., by requiring the continuous deployment pipeline to be used for any change, even a small change in configuration. This avoids errors caused by ad hoc deployments and resulting misconfiguration. The practices also refer to the time that it takes to diagnose and repair an error. The normal deployment process should make it easy to trace the history of a particular virtual machine image and understand the components that were included in that image.

  In general, when a process becomes enforced, some individuals may be required to change their normal operating procedures and, possibly, the structure of the systems on which they work. One point where a deployment process could be enforced is in the initiation phase of each system. Each system, when it is initialized, verifies its pedigree. That is, it arrived at execution through a series of steps, each of which can be checked to have occurred. Furthermore, the systems on which it depends, e.g., operating systems or middleware, also have verifiable pedigrees.

Figure 4.1 gives an overview of DevOps processes.

At its most basic, DevOps advocates treating operations personnel as first-class stakeholders. Preparing a release can be a very serious and onerous process. As such, operations personnel may need to be trained in the types of runtime errors that can occur in a system under development; they may have suggestions as to the type and structure of log files; and they may provide other types of input into the requirements process. At its most extreme, DevOps practices make developers responsible for monitoring the progress and errors that occur during deployment and execution, so

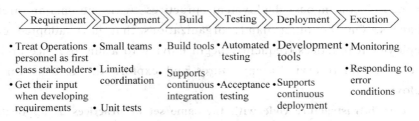

Figure 4.1 DevOps lifecycle processes.

theirs would be the voices suggesting requirements. In between are practices that cover teams, build processes, testing processes, and deployment processes. We discuss continuous deployment, monitoring, security, and more in dedicated chapters of our book.

4.4 Organizational Aspects of DevOps

One main difference between DevOps and traditional models of software development is team size. Although the exact team size recommendation differs from one methodology to another, all agree that the size of the team should be relatively small. Amazon has a *two pizza rule*: No team should be larger than can be fed from two pizzas. Although there is a fair bit of ambiguity in this rule (how big are the pizzas, how hungry are the members of the team), the intent is clear.

The advantages of small teams are:
- They can make decisions quickly. In every meeting, attendees wish to express their opinions. The smaller the number of attendees at the meeting, the fewer the number of opinions expressed and the less time spent hearing differing opinions. Consequently, the opinions can be expressed and a consensus arrived at faster than with a large team.
- It is easier to fashion a small number of people into a coherent unit than a large number. A *coherent unit* is one in which everyone understands and subscribes to a common set of goals for the team.
- It is easier for individuals to express an opinion or idea in front of a small group than in front of a large one.

The disadvantage of a small team is that some tasks are larger than what can be accomplished by a small number of individuals. In this case, the task has to be broken up into smaller pieces, each given to a different team and the different pieces need to work together sufficiently well to accomplish the larger task. To achieve this, the teams need to coordinate. However, coordination needs to be asynchronous and implicit, best achieved through a suitable architecture and interfaces, else all benefits are counterbalanced by the additional coordination overhead.

The team size becomes a major driver of the overall architecture. A small team, by necessity, works on a small amount of code. We will see that an architecture constructed around a collection of *microservices* is a good means to package these small tasks and reduce the need for explicit coordination, so we will call the output of a development team a service. We give an overview of microservice architectures driven by small teams next.

4.5　Implications for Software Architecture: Microservices

DevOps achieves its goals partially by replacing explicit coordination with implicit and often less coordination, and we will see how the architecture of the system being developed acts as the implicit coordination mechanism.

As said previously, development teams using DevOps processes are usually small and should have limited inter-team coordination. Small teams imply that each team has a limited scope in terms of the components they develop. When a team deploys a component, it cannot go into production unless the component is compatible with other components with which it interacts. This compatibility can be ensured explicitly through multi-team coordination, or it can be ensured implicitly through the definition of the architecture.

An organization can introduce continuous deployment without major architectural modifications. Dramatically reducing the time required to place a component into production, however, will require architectural support:
- Deploying without the necessity of explicit coordination with other teams will reduce the time required to place a component into production.
- Allowing for different versions of the same service to be simultaneously in production will allow different team members to deploy without coordination with other members of their team.
- Rolling back a deployment in the event of errors allows for various forms of live testing.

An architectural style that satisfies these requirements is a *microservice architecture*. This style is used in practice by organizations that have adopted or inspired many DevOps practices. Although project requirements may cause deviations to this style, it remains a good general basis for projects that are adopting DevOps practices.

A microservice architecture consists of a collection of services where each service provides a small amount of functionality and the total functionality of the system is derived from composing multiple services. A microservice architecture, with some modifications, gives each team the ability to deploy its service independently from other teams, to have multiple versions of a service in production simultaneously, and to roll back to a prior version relatively easily.

Figure 4.2 describes the situation that results from using a microservice architecture. A user interacts with a single consumer-facing service. This service, in turn, utilizes a collection of other services. We use the terminology "service" to refer to a component that provides a service and "client" to refer to a component that requests a service. A single component can be a client in one interaction and a service in another. In a system such as LinkedIn, the service depth may reach as much as 70 for a single user request.

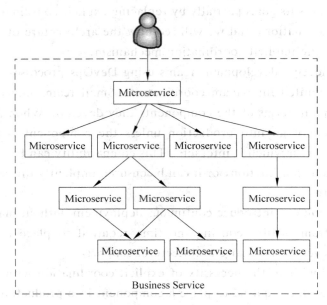

Figure 4.2　User interacting with a single service that, in turn, utilizes multiple other services.

4.6　Summary

The main takeaway from this article is that people have defined DevOps from different perspectives, but one common objective is to reduce the time between a feature or improvement being conceived to its eventual deployment to users, without sacrificing quality.

DevOps will face barriers due to both culture and technical challenges. It will have a huge impact on team structure, software architecture, and traditional ways of conducting operations. We have given you a taste of it by listing some common practices.

The DevOps goal of minimizing coordination among various teams can be achieved by using a microservice architectural style where the coordination mechanism, the resource management decisions, and the mapping of architectural elements are all specified by the architecture and hence require minimal inter-team coordination.

All in all, we believe that DevOps will lead IT onto exciting new ground, with high frequency of innovation and fast cycles to improve the user experience. This article is only a short summary of our book *DevOps*: *A Software Architect's Perspective*. In the book, you can find more information on the topics highlighted here, as well as quality aspects, monitoring practices, three case studies from practice, and much more.

4.7　New Words and Expressions

dependability and availability	可靠性和可用性
delivery mechanism	输送装置;传送机构
goal-oriented	*adj.* 面向目标的
analogously	*adv.* 类似地;近似地
as well	也;同样地
diagnose	*vi.* 诊断;判断
pedigrees	*n.* 家系;谱系
stakeholders	*n.* 利益相关者
onerous	*adj.* 繁重的;麻烦的;烦琐的
ambiguity	*n.* 含糊;不明确
consensus	*n.* 一致;公识
counterbalance	*vt.* 使平衡;抵消
implicitly	*adv.* 含蓄地;暗中地
deviation	*n.* 偏差;误差

4.8　Questions

Single Choice Questions

1. Which kind of the following area of the Ops requirements is not mentioned to support incident handling? (　　)
　　A. Logging　　　B. Modification　　C. Information　　D. Monitoring
2. Which of the following statements is true? (　　)
　　A. Organizations that utilize these practices typically have a period of time in which Dev has primary responsibility for a new deployment all the time
　　B. Organizations that utilize these practices typically have a period of time in

which Ops has primary responsibility for a new deployment all the time

C. Organizations that utilize these practices typically have a period of time in which Dev has primary responsibility for a new deployment; later on Ops has primary responsibility

D. Organizations that utilize these practices typically have a period of time in which Ops has primary responsibility for a new deployment; later on Dev has primary responsibility

3. Which of the following statements is not true? (   )

   A. Once Dev becomes aware of the requirements for incident handling, some architectural modifications may result

   B. Once some architectural modifications may result, Dev becomes aware of the requirements for incident handling

   C. After Dev becomes aware of the requirements for incident handling, some architectural modifications may result

   D. Some architectural modifications may result before Dev becomes aware of the requirements for incident handling

4. Which of the following sentence is not the included in the passage? (   )

   A. Practices associated with continuous deployment are intended to shorten the time between a developer committing code to a repository

   B. Continuous deployment also emphasizes automated tests

   C. Continuous deployment is the practice which leads to the most far-reaching architectural modifications

   D. An organization can introduce continuous deployment practices without any changes

5. Which statement is true according to the passage? (   )

   A. Practices that apply to the development of infrastructure code are intended neither to ensure high quality in the deployed applications nor to ensure that deployments proceed as planned

   B. Errors in deployment scripts such as misconfigurations can cause errors in the application, in the environment, or in the deployment process

   C. Applying quality control practices used in normal software development when developing operations scripts and processes will help control the quality of these specifications

   D. These practices will not affect the application code but may affect the architecture of the infrastructure code

6. What kind of the process can it be for preparing a release? (   )

   A. It can be a very serious and facile process

   B. It can be a very serious and onerous process

C. It can be a very easy and onerous process

D. It can be a very facile and easy process

7. Which of the following statements is the *pizza rule* of Amazon? (　　)

   A. No team should be larger than can be fed from one pizza

   B. No team should be larger than can be fed from two pizzas

   C. No team should be larger than can be fed from three pizzas

   D. No team should be larger than can be fed from four pizzas

8. Which of the following sentences is included according to the passage? (　　)

   A. The larger the number of attendees at the meeting, the fewer the number of opinions expressed and the less time spent hearing differing opinions

   B. The larger the number of attendees at the meeting, the more the number of opinions expressed and the less time spent hearing differing opinions

   C. The smaller the number of attendees at the meeting, the fewer the number of opinions expressed and the less time spent hearing differing opinions

   D. The smaller the number of attendees at the meeting, the more the number of opinions expressed and the less time spent hearing differing opinions

9. Which of the following statements is true? (　　)

   A. Allowing for different versions of the same service to be simultaneously in production will allow different team members to deploy without coordination with other members of their team

   B. Allowing for different versions of the same service to be simultaneously in production will allow different team members to deploy with coordination with other members of their team

   C. Allowing for different versions of the same service to be asynchronously in production will allow different team members to deploy without coordination with other members of their team

   D. Allowing for different versions of the same service to be asynchronously in production will allow different team members to deploy with coordination with other members of their team

10. Which of the following functions can not be implemented in a microservice architecture? (　　)

    A. Gives each team the ability to deploy its service independently from other teams

    B. Having multiple versions of a service in production simultaneously

    C. Rolling back to a prior version relatively easily

    D. Utilizing a collection of other services

4.9　Problems

After reading this chapter and completing the exercises, you will be able to do the following:

1. What Software Architects Need to Know About DevOps?
2. Explain the defining DevOps.
3. Discuss DevOps Practices and Architectural Implications.
4. Explain an overview of DevOps processes.
5. Discuss Organizational Aspects of Dev.
6. Discuss Implications for Software Architecture: Microservices.

Unit 5 Office Computing—The Ultimate Player's Guide to Minecraft-Xbox Edition: Gathering Resources

This chapter is about building the foundation you can use to launch into the rest of the game. Your focus is on a few key points: build an outdoor shelter, find food to stave off hunger, improve your collection of tools, and build a chest to safely store items.

Minecraft is filled to the brim with all manner of resources, and gathering them is the first step toward getting the most out of the game. This chapter is about building the foundation you can use to launch into the rest of the game. Your focus is on a few key points: build an outdoor shelter, find food to stave off hunger, improve your collection of tools, and build a chest to safely store items. This solidifies your position (making your base more impervious to attack), allows you to do all sorts of Minecrafty things more efficiently, and sets you up for longer excursions both above and below ground.

The good news is that you already have a base, so you can explore during the day, try not to lose your way, and head back at night. However, you still need to avoid at least some of the hostile mobs that persist during the day.

5.1 Introducing the HUD

Before we start, let's take a look at the Heads-Up Display (HUD)—that collection of icons and status bars at the bottom of the screen. Figure 5.1 shows the HUD as it appears in Survival mode with all possible indicators displayed. (The Creative mode HUD shows only the Inventory bar.)

Figure 5.1 The HUD provides key status indications.

Health is all important, but low hunger also leads to low health, so keep a close eye on both.

1. Armor bar
2. Health bar
3. Experience bar
4. Oxygen bar
5. Hunger bar
6. Hotbar

Each section of the HUD provides a key nugget of information about the health or status of your avatar:

- **Armor bar**—The armor bar appears when you've equipped your avatar with any type of armor and shows the current damage absorption level. Each armor icon represents an 8% reduction in the damage you'll take, so a 10/10 suit of armor reduces the damage you take by 80%, whereas a 1/10 suit absorbs only 8%. Armor becomes less effective the more damage it has also taken, although the rate at which it deteriorates also depends on its material—leather being the weakest and diamond the strongest. In the case shown in Figure 5.1, a set of leather boots really doesn't provide much protection.

- **Health bar**—You have up to 20 points of health available, represented by the 10 hearts shown. Each heart disappears in two ticks. Health and hunger have a complicated relationship. You can read more about them starting in the section "Hunger Management."

- **Experience bar**—The experience bar increases the more you mine, smelt, cook, kill, and fish. Your current level is shown in the middle of the bar. When it's full, you move to the next experience level. Experience isn't generally important until you start enchanting and giving additional powers to items such as swords. Unlike other role-playing games, experience in Minecraft is more like a currency that you spend on enchantments, so it waxes and wanes. But all experience gained counts toward the final score shown on the screen when you die. Killing a mob drops experience orbs that either fly directly toward you or float to the ground waiting to be collected. You can also gain experience by smelting certain items in the furnace and carrying out other activities such as finding rare ores. Dying causes a substantial drop in your current experience level, so if you start to gain substantial experience points (for example, a level that's up in the 20s), it might be time to think about spending them on an enchantment or two.

- **Oxygen bar**—The oxygen bar appears whenever you go underwater, and it quickly starts to drop. You can probably hold your own breath for longer! As soon as your oxygen level hits zero, your health starts taking a two-point hit every second, but it resets if you resurface for just an instant. You can do this by holding down the jump key until you breach the water. Diving isn't that big of a deal in Minecraft, at least not for completing the core game, but you can use the ability to do interesting things like building an underwater base. An example is shown in Figure 5.2, "Creative Construction", as well as sharing with you some other underwater breathing techniques.

Figure 5.2 Creative Constructions

Underwater bases are impervious to mob attacks, even when built from glass, but you'll need to watch your oxygen bar carefully to ensure you don't run out of air while building this type of structure. By the way, the only mob that spawns underwater is the friendly, curious squid. Can you make out the one shown here? Say hello to Ceph.

- **Hunger bar**—You also have 20 points of hunger available, as well as a hidden value called Saturation. Like health, each hunger bar icon holds 2 points and can reduce by half an icon (that icon is, incidentally, a "shank," or the lower part of a leg of meat) at a time.
- **Hotbar**—These nine slots represent items you can select and use. Press (Y) to access your full inventory and to change the items in these slots. The white number next to some shows that slot's count of stacked identical items. A durability bar also appears under each tool's icon in green, gradually reducing as you use them until the tool breaks and disappears from your inventory. You'll have some warning of this because the bar turns red when it's close to zero. See "Improving Your Tools" later in the chapter to learn more about the durability of various materials.

NOTE: Hiding the HUD

If you want to hide the HUD, press (▶) to open the **Help & Options** menu. Scroll down to **Settings** → **User Interface** and deselect **Display HUD**. Unfortunately, there isn't a quicker way to do this at present.

5.2　Avoiding Getting Lost

It's easy to become lost in Minecraft. Run helter-skelter from your base, chase a herd of livestock, discover a natural cave system, or take a shot across the sea like that famed Norseman, Leif Eriksson. It's all part of the Minecraft charm. But don't become Columbus in the process.

You'll find a map in your inventory that can help you always return to your home base or other locations in the world (see Figure 5.3). The map can display the entire world but only updates while you have it active in the Hotbar, so it will take some time for it to build up the big picture. However, it does provide coordinates. Take note of those displayed for your home base.

Figure 5.3　Point your (RS) down to view the map. In this screenshot I've also turned off the HUD for a better view.

The coordinates are based on the world's center where X and Z equal 0. (Y shows your current level above bedrock.) Jot down the current values. If you become lost, you can always find your way back to your original spawn and, presumably, your first shelter by traveling in a direction that will bring both X and Z back to those noted values.

When you need to return—and I should warn you that this *can* take some experimentation and a little practice—turn and take a few steps while tracking the change in your current coordinates. Your goal is to shift those X and Z values back toward those you originally recorded. You'll probably wander around a bit, but eventually you'll get there and the map will help you get your orientation and to head off in the right general direction.

When you are able, craft a compass. It takes some redstone and iron, and both are relatively easy to obtain with some assiduous mining. The only problem with a compass is that it always points to your original world spawn point. Think of that point as the magnetic north pole—it's not a GPS. Sleeping in a bed resets your spawn point but not your compass, so this method falls out of date as soon as you move to new dwellings and update your spawn point.

A compass is actually more useful when transformed into a map. You may need to do that if you lose the original map.

5.3 Improving Your Tools

Wooden tools wear out fast, so it's best to upgrade your kit as quickly as possible.

Each type of material has a different level of *durability*. Think of durability as the number of useful actions the tool can perform before it wears out completely, disappearing in a sudden splintering of wood. I've included the durability in parentheses after each material's description:

- **Gold (33)**—Although this is the least durable material, a gold pickaxe can break blocks out of most softer materials in the blink of an eye, and it happens to be the most enchantable material, so you can imbue it with superpowers. But given that gold is about five times as rare as iron, and gold can be used to craft many other useful items, I wouldn't recommend using it for tools.
- **Wood (60)**—It's easy to obtain, especially in an emergency aboveground, but think of wood as just a means of getting to cobblestone because, unlike the latter, wooden tools can't mine the more valuable ores such as iron, gold, diamond, and redstone. You will at least need a wooden pickaxe to mine stone because doing so with your bare hands will just break the stone into unusable dust, but after that, swap them out for something tougher.
- **Stone (132)**—With just a touch over twice the longevity of wood, stone makes a great starting point for more serious mining and other activities such as slaying mobs. Stone tools are built from cobblestone blocks, which in turn come from stone. That may seem a little confusing, but it will seem natural enough after a while.
- **Iron (251)**—Iron will become your go-to material. It is found most commonly all the way from bedrock, the lowest layer of the Minecraft world, up to about 20 levels below sea level. Iron is used for building all kinds of tools, implements, and devices including armor, buckets (for carrying water, lava, and milk), compasses, minecarts, and minecart tracks. All these require at least iron ingots obtained by smelting the ore in a furnace, with each block of

ore producing one ingot. Ingots and many other items are found scattered throughout the world in village chests, mine shafts, dungeons, and strongholds. You might also find them as drops from slain zombies and iron golems—although I definitely don't recommend tackling the latter.

- **Diamond** (1562)—It's the strongest material of all but also the most expensive given that diamonds are relatively rare. (You will enjoy the moment you do find your first diamond, but it's found only in the first 16 layers above bedrock, the lowest layer in the Minecraft world, and even then it's about 25 times as scarce as iron.) A diamond pickaxe is the only material that can successfully mine obsidian, a material required for creating the portal to reach The Nether region. Given that diamond is much scarcer than iron but only six times as durable, you should use iron pickaxes as much as possible and only switch to diamond when you need to mine obsidian to reach The Nether. You're better off saving any diamonds you find for weapons (a diamond sword does more damage, and that combined with its increased durability will ensure it lasts much longer than any other material while doing more damage where it counts), armor, and enchantment tables.

NOTE: Different Materials for Different Items

Durability applies to all tools, weapons, and armor, although there are differences in the materials that can be used in each case. For example, you can craft leather armor and make stone tools, but not vice versa.

The recipes for crafting tools from all materials are identical, save for the replacement of the head of the implement with the material of choice. As long as you have the right materials, that version of the tool appears selectable in the crafting interface:

- To make a stone pick, you need two wooden sticks for the handle and three cobblestone blocks.
- Replace in the same way for the axe and the sword.
- You might also want to add a shovel to your collection because it's about four times faster than using hands to harvest softer materials such as dirt, gravel, sand, clay, and snow, and it helps some of those blocks deliver resources rather than just breaking down. For example, only a shovel can gather snow balls from snow.

As you craft more items, you need to find somewhere to store those you don't need to use right away. You should also store other resources and food you find on your travels so they're not lost if you come to an unfortunate end. That comes next.

5.4 Chests: Safely Stashing Your Stuff

Whenever you head away from your secure shelter, there is always a reasonably high risk of death. Creepers, lava pits, long falls—they can all do you in. Respawning is only a moment away, but the real danger here is that any items you've collected and carry in your character's inventory drop at the location of your untimely death and may prove impossible to retrieve in the 5 minutes you have to get back to them before they disappear forever.

Chests act as an insurance policy. Put everything you don't need in a chest before you embark on a mission, and those things will be there when you get back or after you respawn.

The natural place to leave chests is in your shelter, but you can also leave them elsewhere, perhaps at a staging point as you work away in a mine or even outside. Mobs will leave them alone, and the only real risk you face is leaving them out in the open in a multiplayer game, or getting blown up from behind by a creeper in Single-player mode while you're rummaging around inside.

Chests come in two sizes: single and double. A single chest can store 27 stacks of items. Create a double chest by placing two single chests side by side. The double chest stores up to 54 stacks of items. Given that a stack can be up to 64 items high, that's an astonishing potential total of 3,510 blocks in a crate that takes just 2×1 blocks of floor space. If you've ever followed the *Doctor Who* TV series, consider chests the Tardis of storage!

Create a chest at your crafting table with eight blocks of wooden planks.

Place and then use [LT] to open it. You can then move items back and forth between your inventory and the chest. In Figure 5.4, I've transferred all the items I don't need for the next expedition.

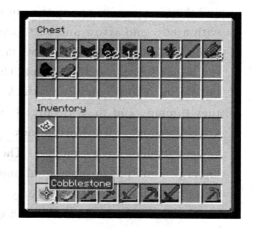

Figure 5.4 Chests act as an insurance policy for your items so they aren't lost if you die. Use the inventory shortcuts you learned earlier to quickly move items between your active inventory and the chest's storage slots.

Before you head out, there are two other things you should know: how to avoid monsters and how to deal with hunger. Read on.

5.5 Avoiding Monsters

There's a key difference between the Minecraft world on the first and second days. In a word, *mobs*: hostile ones to be specific. Mobs only spawn in dark areas, and some only during the night, so if you are outside during your first day and stay in well-lit areas, you'll be reasonably—although not entirely—safe. By the second day, however, mobs have had a chance to build their numbers and wander about. It's not that likely you'll encounter them on day 2, but it's best to be prepared.

There are 14 types of hostile mobs. These are the ones you might meet on your second day outside:

- **Zombies**—Zombies burn up in sunlight but can still survive in shadows or rain, when wearing a helmet, and of course in caves all hours of the day or night. They are relatively easy to defeat, and if too many come after you from out of the shadows, just head to a well-lit area and keep your distance while they burn up in the sun.
- **Skeletons**—Skeletons also burn up in sunlight unless they are wearing a helmet, and they can survive at any time in lower light conditions. They're quite deadly with a bow and arrow and best avoided for now.
- **Spiders**—Spiders come in two variants: large or blue. You'll probably only see the larger spiders at this stage. They are passive during the day but become hostile in shadow and can attack at any time if provoked. They'll climb, they'll jump, and they are pretty darn fast. Fortunately, they're also easy to kill with some swift sword attacks. The blue spiders are a smaller, poisonous variant called cave spiders. They live only in abandoned mine shafts underground, but in substantial numbers. If you suffer from arachnophobia, I don't have much good news for you, except that with a little time you'll get used to them and they won't seem quite so nasty.
- **Creepers**—Creepers have a well-earned reputation as the Minecraft bad guys. They are packed to their green gills with gunpowder, and they'll start their very short 1.5-second fuse as soon as they are within three blocks of you. Their explosions can cause a lot of real damage to you, nearby structures, and the environment in general. If you hear a creeper's fuse—a soft hissing noise—but can't see it, run like heck in the direction you're facing. Remember to sprint by pushing your Ⓛ forward twice in quick succession. With a little luck, you'll get three blocks away and the creeper's fuse will reset. Creepers are usually best dealt with using a ranged attack from a bow and arrow, but if you sprint at them with an iron or diamond sword and take a swipe at just the right

Unit 5 Office Computing—The Ultimate Player's Guide to Minecraft-Xbox Edition: Gathering Resources

moment, mid-leap, when you're past the apex of the arc and descending in a wild fury, you can send them flying back out of their suicidal detonation range, causing the fuse to reset. Most creepers despawn (that is, disappear) around noon, leaving the afternoon generally free of their particular brand of terror.

- **Slimes**—Slimes spawn in the swamp biome and in some places underground. Their initial appearance is that of a quite large Jello-like green block, but they won't sit there gently shaking; they are more than capable of causing real damage. Attacking eventually breaks them up into 2-4 new medium-sized slimes. These can still attack but are relatively easily killed, only to spawn a further 2-4 tiny slimes each! These last don't cause any attack damage, but may still push you into peril if you're unlucky.

If you come across a lone spider, a zombie, or even a slime, now is as good a time as any to get in some sword practice. Just point your crosshairs at the creature and strike repeatedly with [RT]. Timed well, you can also block their attacks with [LT]. Keep clicking as fast as you can, and you've got a very good chance of killing the mob and picking up any items it drops before it lands too many blows. Try to avoid the other mobs for now.

TIP: Switch to Peaceful Mode to Get a Break

Getting mobbed by mobs? Remember that you can always exit the game and reenter it, changing the Difficulty setting to Peaceful as you open the game once more. Peaceful mode despawns all hostile mobs and allows your health to regenerate. But do try to switch the level back to Normal as soon as you can.

So how do you avoid mobs? Use these tips to survive:

- Stay in the open as much as you can, avoiding heavily wooded areas if possible.
- Most mobs have a 16-block detection radar. If they can also draw a line of sight to your position, they will enter *pursuit* mode. (Spiders can always detect you, even though other blocks.) At that point they'll relentlessly plot and follow a path to your position, tracking you through other blocks without requiring a line of sight. Pursuit mode stays engaged much farther than 16 blocks.
- Keep your sound turned up because you'll also hear mobs within 16 blocks, although creepers, befitting their name, are creepily quiet.
- Avoid skirting along the edges of hilly terrain. Creepers can drop on you from above with their fuses already ticking. Try to head directly up and down hills so you have a good view of the terrain ahead.
- Mobs are quite slow, so you can easily put some distance between them and

yourself by keeping up a steady pace and circling around to get back to your shelter. Sprint mode will leave them far behind.

CAUTION: Sprinting Makes You Hungry

Sprint mode burns up hunger points, so try to use it only in emergencies. Unfortunately, in a real emergency, making a dash from a creeper when your health is low, you'll find it impossible to sprint. Remember, always keep your hunger topped up so your health continually regenerates and you'll avoid getting caught in this leaden-footed nightmare.

5.6 Hunger Management

Hunger plays a permanent role in Minecraft, much as in real life. While it's only possible to starve to death on Hard difficulty, hunger does affect your character in other ways, so it's always important to ensure you have the equivalent of a couple of sandwiches packed before heading deep into a mine or on a long trek.

Hunger is a combination of two values: the one shown in the HUD's hunger bar, as well as a hidden value called *saturation*. The latter provides a buffer to the hunger bar, decreasing first. In fact, your hunger bar doesn't decrease at all until saturation reaches 0. At that point, you see the hunger bar start to jitter, and after a short while it takes its first hit. Saturation cannot exceed the value of the hunger bar, so with a fully satiated bar of 20 points, it's possible to have up to 20 points of saturation. However, a hunger level of 6 points also only provides a maximum of 6 points of saturation, and that makes you vulnerable.

You'll find some key information about the hunger system here:

- On Easy and Normal Survival modes, your character won't drop dead from hunger, although it can still pose a danger because your health won't regenerate if hunger has dropped more than 2 points from its maximum. If you're close to home and pottering around in your farm or constructing some building extensions, you're fairly safe, but your health *will* start to drop. Eat something as soon as you can to rebuild your hunger bar and therefore your health.
- Sprinting and jumping up blocks both cost hunger points. Also, sprinting becomes impossible when the hunger bar drops below 6 hunger points, or 3 shanks, as shown in the HUD.
- Keeping a relatively full stomach at 18 hunger points (9 shanks in the HUD) allows health to regenerate at 1 point (half a heart) every 4 seconds.
- Health depletes if the hunger bar drops to 0, increasing the risk of dying from one of the many imaginative ways on offer in Minecraft's deadly smorgasbord

(see Figure 5.5).

Figure 5.5 The effects of extreme hunger on Normal difficulty:
health depletes to just one point, or half a heart.

- There are some limits to the amount health can drop that vary according to the difficulty level. On Easy, health cannot deplete from hunger further than 10 points, or half the full quotient. On Normal, it drops to 1 point, which is an extreme level of vulnerability. On Hard, there are no limits; don't ignore the hunger bar, or death from starvation could be just moments away. See "Food on the Run" later in this chapter to avoid this unfortunate fate.

5.7 Your Mission: Food, Resources, and Reconnaissance

Your second day is the perfect opportunity to gather food and other resources and to take a quick survey of the landscape surrounding your first shelter, in particular to find somewhere suitable for your first outdoor abode. Keep an eye out for any of the following:

- **Passive mobs**—Chickens, pigs, and cows all provide a ready source of food, or at least raw protein that can be cooked on the furnace and made more nutritious. Cows also drop leather that you can use for your first armor and can be milked, giving you an instant cure for food poisoning. Chickens also lay eggs, which are used to make cake, so gather any that you find. You can also obtain feathers from killed chickens—useful for later crafting arrows.
- **Natural harvest**—The harvest includes cocoa pods, apples, sugar cane, carrots, potatoes, and seeds. Knock down tall grass to find seeds; then use a hoe to till some ground next to water. Seeds mature into wheat within 5 – 8 day/night cycles, although wheat is also grown by villagers, as you can see with the wheat crop shown in Figure 5.6. From wheat, it's easy to bake bread, one of the simplest but most effective sources of food, especially if there are no passive mobs nearby. When combined with cocoa pods, bread will make cookies, which are always useful for a quick hunger bar top-up.

Figure 5.6 Wheat is an easy crop to farm and then to turn into bread—
a handy food if you're stuck with no other options.

- **Construction resources**—You can mine plenty of cobblestone quite safely by expanding your original shelter, digging into the terrain. But some other resources will definitely come in handy. Wood is always useful. If you see any sand, mine it so you can smelt it into glass blocks to let light into your shelter and provide a view. (There's no point moving from your first cave into the outdoor equivalent of another!) Also keep an eye out for coal. You can often see it in veins on the surface of the walls of small caves or on the sides of cliffs. If you can safely get to it, make like a miner and dig it out. Use the coal to make torches and to smelt other ores.

TIP: Making Use of Bones

The morning sun burns up skeletons, leaving behind bones that you can craft into bone meal. Bone meal acts as a fertilizer, helping your crops grow faster. You can also use bone to tame wolves, providing you with an extra level of protection.

Start early, heading out with a stone sword at the ready, just in case. If you are low on wood, swing an axe at a few nearby trees.

Move carefully so you don't lose your bearings. The sun rises in the east and sets in the west, and the clouds also travel from east to west, so you can always at least get your bearings. Following a compass cardinal point (north, south, east, or west) using the sun and clouds as a reference can lead you away and reasonably accurately back home again.

5.7.1 Food on the Run

If you are getting dangerously hungry, head to the nearest equivalent of a fast food outlet—a passive mob—sword at the ready. Your best bet is to look for cows and pigs because they each can drop up to three pieces of raw meat, with each restoring 3 hunger units and 1.8 in saturation. They're an excellent target of opportunity. You can also eat raw chicken, although with a 30% chance of developing food poisoning, or you can try rotten meat harvested from zombies, which is guaranteed to give you a taste of the stomach aches. But after you have mined three pieces of iron and crafted a bucket, you can also cure any type of food poisoning by drinking milk obtained with that bucket from a cow. You can also eat any amount of poisoned meat, gaining the restorative benefits, and curing the whole lot with one serving of milk. So keep that rotten flesh the zombies drop around! And the bucket o' milk.

That said, unless you are desperate, it is actually much better to take the time to cook all your meat first. In fact, the secondary processing of foods makes them all healthier, restoring more hunger and saturation points. If you are far from home you could choose to always carry a furnace in your inventory, along with fuel. Place it, cook, and break it up to use again. Or you could, if you don't mind seeming like a crazed pyromaniac, both kill and cook pigs, chickens, and cows in one blazing swoop by setting the ground beneath them on fire with a flint and steel (click on the ground next to the animal) or, a little more chaotically, by pouring lava from a bucket. Just take caution that you don't do this anywhere it could pose a risk such as near that fantastic wood cabin you just spent the last three weeks building; there's no undo in Minecraft.

NOTE: Fishing in the Sea of Plenty

Mobs such as chickens, cows, and pigs spawn quite rarely compared to hostile mobs, so consider them a nonrenewable resource if you kill them in the wild. You're better off breeding them in a farm so they can be readily replaced. Fish, on the other hand, are unlimited in quantity and very plentiful, and fishing from a boat works very well. Your hunger bar won't decrease, and you'll be relatively safe from hostile mobs. Even better, you can eat on the go as you won't ever get food poisoning from raw fish. Sushi anyone?

TIP: Let Them Eat Cake

What's the quickest way to fill your hunger bar? Eat cake. Unlike another well-known game, Minecraft's cake is not a lie. Cake has a quite complicated recipe, but each full cake provides up to 6 slices, each worth 1.5 hunger points, or 9 in total, and it's less resource intensive than creating golden apples. Minecraft rewards calories, so eat as much as you like without penalty, quickly building back your full hunger bar but, as in the real world, the nutrients are lacking, so cake doesn't provide any saturation benefit. Make sure you eat some more nutritional foods such as protein as your hunger bar starts to top out to ensure you also get that extra boost. Personally speaking, if there was a choice between cake and pizza, I'm going for the latter!

Finally, if you simply cannot find mobs, your hunger bar has dropped to starvation, and your health has plummeted to half a point, consider at least planting a wheat field and waiting it out in your shelter for a few days so that at least three blocks of wheat can grow and be baked into bread.

If all is lost, even then, consider one final alternative—a pretty neat if somewhat dramatic trick. Assuming you have reset your spawn point to a bed or are still near origin, head to your shelter, place everything you carry in a chest, and then head outside and either jump off a cliff, drown in a lake, or wait for a mob to kill you. You respawn back in your shelter with full health, a full hunger bar, and all your possessions waiting for you in the chest. The only downside is that you'll lose some experience points in the process, which impacts enchanting, but I'm sure you can build those up again quick enough. It's a good last resort, and will let you quickly head out again, fully equipped, to live another day.

5.7.2 Finding a Building Site

As you scout around, keep in mind that you are also looking for a new building site. This doesn't have to be fancy or even particularly large. A 6×5 space manages just fine, and even 6×4 can squeeze in the basics. You can also level ground and break down a few trees to clear space. I did this in Figure 5.7.

Figure 5.7 A nice, flat, elevated building site created on a nearby hill after filling out the platform with dirt.

I usually prefer space that's a little elevated because it provides a better view of the surroundings, but it's perfectly possible to create a protected space just about anywhere. You may even decide to go a little hybrid, building a house that's both tunneled into a hill and extending outside.

TIP: Light Those Caves

Check for any caves or tunnels close to your site's location. If they aren't too big, light them up with torches to prevent mobs spawning inside and wandering out during the day, or just block their entrances for now.

So what can we build on this site? Let me show you a basic structure. It takes 34 cobblestone blocks dug out of the first shelter and 12 wood blocks for the roof obtained by cutting down 3 trees. This is about as minimalistic as it gets.

You can build the roof from almost any handy material, including dirt, cobblestone, and wood, as shown in Figure 5.8. Avoid blocks that fall down, such as gravel and sand. A two-block-high wall keeps out all mobs except for spiders because they can climb walls. An overhang on the wall keeps spiders out because they can't climb upside-down, but it's easier to just add a roof, and this will protect you if there are any trees nearby the spiders can climb and use as an arachnid's springboard to jump straight into your dwelling. (Yes, it's happened to me. Sent shivers up my spine.) Figure 5.9 shows the finished hut with a few torches on the outside to keep things well lit.

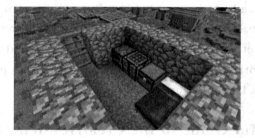

Figure 5.8   The layout for a small cobblestone cabin using a total of 46 blocks, roof not shown. The sharp-eyed will notice it can be reduced in width one space further, but for the sake of four blocks, that would feel just a little too claustrophobic.

TIP: No Housing Codes in Minecraft

The roof in Figure 5.9 rests right on the lip of the inner wall. You can't directly build a roof like this from scratch. First, place a block on top of the wall, and then attach the inner block for the roof. Remove the first block, and the inner block floats. Attach new blocks to that to build out the roof structure. It won't pass a building inspection, but it certainly works in Minecraft.

Figure 5.9   The finished hut—basic but serviceable. And it's spider proof. Although there is a large gap above the door, in Minecraft's geometry the door fills the entire space. Spiders are two blocks wide, so they can't fit through a one-block-wide gap. You could actually leave the door wide open, and spiders will just gather outside and make horrible noises, but don't do that because it's an invitation for other mobs to enter.

Building a wall even two blocks high can take a little bit of fancy footwork. Some basic techniques help:

- Place your walls one layer at time. Put down the first layer, and then jump on top to place the second.
- If you fall off, place a temporary block on the inside of your structure against the wall, and use this to climb back up. You can remove it when you're finished.
- Use pillar jumping if you need to go higher. While looking directly down, press Ⓐ to jump and then use ⓇⓉ to place a block underneath you. You land on that block instead of the one below. Repeat as often as necessary. Dig the blocks out from directly underneath you to go back down.
- Click ⓇⓈ to activate Sneak mode as you work around the top of tall walls so you don't fall off. You can even use this technique to place blocks on the side of your current layer that are normally beyond sight.

5.8 A Resourceful Guide to the Creative Mode Inventory

Minecraft resources fall into several primary categories. Some of them are a natural early focus as you improve your position from those gathered for first-night survival; others come into more focus as you get further through the game, gear up for your exploration of The Nether and The End regions, and start to become more creative with all that Minecraft has to offer. Here's a quick summary of the categories. You can view all the possible tools and resources by opening a game in Creative mode and pressing Ⓧ, as shown in Figure 5.10. The categories that follow correspond to the tabs running across the upper and lower sections of the Creative mode inventory. Scroll the inventory with ⓇⓈ.

- **Building Blocks**—Building blocks are used, as you might expect, for construction, including housing and almost anything else. Build a bridge for your redstone rail. Construct a dam. Elevate a farm above a level that won't get trampled by mobs, or put up a fence. Build a skyscraper or reconstruct a monument. Minecraft provides a large number of primary blocks—cobblestone, gravel, wood, and dirt, etc.—that can be harvested directly, but things definitely become more interesting once you start creating secondary types of blocks from primary materials. You can store many items more efficiently (for example, by converting nine gold ingots into a single gold block) and climb more efficiently by crafting stairs instead of jumping up and down blocks on well-travelled routes. These blocks are, without being too punny, the building blocks of creativity.

Unit 5 Office Computing—The Ultimate Player's Guide to Minecraft-Xbox Edition: Gathering Resources

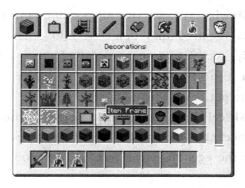

Figure 5.10 Creative mode inventory provides access to the full set of resources and tools.

- **Decorations**—Decorations are something of a catchall category. Generally, they are things you can use to make your constructions more interesting. Some of those are just visual, such as the various mob heads, whereas others, such as item frames and bookcases, also serve functional purposes.
- **Redstone and Transportation**—Redstone is an almost magical resource. You can use it to build powered circuits, quite complex ones, and then activate pistons to automatically harvest a farm plot, set up traps, open and close doors, and a huge amount more. The limits are set only by your imagination. Redstone is also used to craft powered rail tracks and a range of other useful items such as a compass and clock. This category also includes other items used for transportation such as the various types of minecarts and boats. There are enough options here to enable you to build everything from massive transportation systems to incredible rollercoasters.
- **Materials**—Materials is a catchall category, composed of items derived from another action. For example, killing a chicken can drop feathers, and you'll need those for the fletching on arrows unless you gather arrows directly from slain skeletons.
- **Food**—Food contains the full range of edibles, including the enchanted form of the golden apple, the rarest edible in the game. Take a few of these with you the next time you think you'll be in a tight spot, and you might just be able to make it through that moaning zombie horde.
- **Tools, Weapons, and Armor**—Tools can be wielded as weapons, but not very effectively. They are, however, great at digging, chopping, hoeing, and setting a Nether Portal on fire with the flint and steel. You'll also find shears for stripping the wool from sheep, a fishing rod, and the full set of armor and tools.

- **Brewing**—The Brewing tab contains all possible potions and a number of the rarer ingredients required that don't fit into other categories. Potions are incredibly handy, delivering such useful effects as protection from fire—something of an advantage when traveling to The Nether. Use [LT] in this tab to cycle through the potions of different strengths.
- **Miscellaneous**—Miscellaneous contains a range of useful and obscure items. You'll find the buckets quite handy for setting up new water and lava sources, and you can use the eggs to spawn most of the mobs, populating a farm and more.

Use [A] to take individual items, or [Y] to take the full permissible stack. Get rid of a single stack from your Hotbar by picking it up, dragging it off the side of the inventory screen, and pressing [A] once more to drop them. You can also replace items by dropping the new one on top of the old.

5.9 The Bottom Line

Congratulations! You've now learned everything you need to know to understand how your character is doing, improve your tools for better longevity, hopefully not get lost on your travels, and create your first mob-proof outdoor shelter.

These are the keys to Minecraft. Just remember to head back to your chest often to store the valuables you've gathered or to build other chests further afield.

You might also want to consider building a pillar and platform on top of your new shelter. It can help you survey your terrain and act as an easy-to-see landmark when you're out and about. Put some torches on top because mobs can spawn on any platform, no matter how small, and you don't want to poke your head up through the platform only to discover a creeper on a short fuse. It will also help you spot home from a distance.

5.10 New Words and Expressions

ultimate	*adj.* 最终的;极限的
stave off	避开;延缓
brim	*vt.* 使…满溢
impervious	*adj.* 不受影响的,无动于衷的
minecraft	*n.* 我的世界
incidentally	*adv.* 顺便;附带地
shank	*n.* 柄;小腿
inventory	*n.* 存货清单;财产清册

coordinate	*n.* 坐标
presumably	*adv.* 大概；可假定
compass	*n.* 指南针；圆规
durability	*n.* 耐久性；坚固
splinter	*v.* 分裂；冻离
ingot	*n.* 锭；铸块
scattered	*adj.* 分散的；散乱的
dungeon	*n.* 地牢；土牢
recipes	*n.* 食谱；方法
vice versa	反之亦然
embark on	从事，着手；登上船
respawn	*vt.* 复位；*vi.* 重生
expedition	*n.* 远征；探险队
reconnaissance	*n.* 侦察；勘测
resourceful	*adj.* 资源丰富的；足智多谋的

5.11 Questions

Single Choice Questions

1. In what color does a durability bar also appear under each tool's icon, gradually reducing as you use them until the tool breaks and disappears from your inventory? (　　)

　　A. Red　　　　B. Yellow　　　　C. Green　　　　D. Blue

2. Which of the following approaches is not included to become lost in Minecraft? (　　)

　　A. Become Columbus

　　B. Run helter-skelter from your base

　　C. Chase a herd of livestock

　　D. Take a shot across the sea

3. How are the coordinates based on the world's center? (　　)

　　A. where X and Y equal 0

　　B. where X and Z equal 0

　　C. where Y and Z equal 0

　　D. where X, Y and Z equal 0

4. How many blocks of wooden planks do you create a chest at your crafting table with in this passage? (　　)

　　A. Five　　　　B. Six　　　　C. Seven　　　　D. Eight

5. Which of the following tips is not included to avoid mobs? (　　)

A. Stay in the open

B. Keep your sound turned up

C. Avoid skirting along the edges of hilly terrain

D. Stay in hungry mode

6. Which of the following statements is true? (　　)

   A. If you're close to home and pottering around in your farm or constructing some building extensions, you're fairly dangerous, but your health will start to drop

   B. If you're close to home and pottering around in your farm or constructing some building extensions, you're fairly safe, but your health will start to drop

   C. If you're close to home and pottering around in your farm or constructing some building extensions, you're fairly dangerous, but your health will start to rise

   D. If you're close to home and pottering around in your farm or constructing some building extensions, you're fairly safe, but your health will start to rise

7. Which of the following is not the included to keep an eye out? (　　)

   A. Chickens　　　　　　　　B. Cocoa pods

   C. Coal　　　　　　　　　　D. Chocolate

8. What kind of animals can you use bone to tame to provide you with an extra level of protection? (　　)

   A. Lions　　　　　　　　　B. Tigers

   C. Wolves　　　　　　　　D. Leopards

9. In this passage, if somebody is getting dangerously hungry, head to the nearest equivalent of a fast food outlet—a passive mob, what kind of weapon is at the ready? (　　)

   A. Sword　　　B. Spear　　　C. Dagger　　　D. Axe

10. Which of the following statements is true? (　　)

    A. Put some explosive on top because mobs can spawn on any platform, no matter how small

    B. Put some torches on top because mobs can spawn on any platform, no matter how small

    C. Put some bombs on top because mobs can spawn on any platform, no matter how small

    D. Put some poison on top because mobs can spawn on any platform, no matter how small

5.12 Problems

After reading this chapter and completing the exercises, you will be able to do the following:

1. How to learn the secrets of the HUD?
2. How to improve your tools with more durable materials?
3. How to safely store your hard-earned resources?
4. How to learn the easy way to manage hunger?
5. How to build your first outdoor shelter and enjoy the view?
6. How to access the full Creative mode inventory?

Unit 6 Networking—Troubleshooting Methods for Cisco IP Networks

This chapter defines troubleshooting and troubleshooting principles. Next, six different troubleshooting approaches are described. The third section of this chapter presents a troubleshooting example based on each of the six troubleshooting approaches.

This chapter covers the following topics:
- Troubleshooting principles
- Common troubleshooting approaches
- Troubleshooting example using six different approaches

Most modern enterprises depend heavily on the smooth operation of their network infrastructure. Network downtime usually translates to loss of productivity, revenue, and reputation. Network troubleshooting is therefore one of the essential responsibilities of the network support group. The more efficiently and effectively the network support personnel diagnose and resolve problems, the lower impact and damages will be to business. In complex environments, troubleshooting can be a daunting task, and the recommended way to diagnose and resolve problems quickly and effectively is by following a structured approach. Structured network troubleshooting requires well-defined and documented troubleshooting procedures.

This chapter defines troubleshooting and troubleshooting principles. Next, six different troubleshooting approaches are described. The third section of this chapter presents a troubleshooting example based on each of the six troubleshooting approaches.

6.1 Troubleshooting Principles

Troubleshooting is the process that leads to the diagnosis and, if possible, resolution of a problem. Troubleshooting is usually triggered when a person reports a problem. In modern and sophisticated environments that deploy proactive network monitoring tools and techniques, a failure/problem may be discovered and even fixed/resolved before end users notice or business applications get affected by it.

Some people say that a problem does not exist until it is noticed, perceived as a problem, and reported as a problem. This implies that you need to differentiate between a problem, as experienced by the user, and the actual cause of that problem.

The time a problem is reported is not necessarily the same time at which the event causing the problem happened. Also, the reporting user generally equates the problem to the symptoms, whereas the troubleshooter often equates the problem to the root cause. For example, if the Internet connection fails on Saturday in a small company, it is usually not a problem, but you can be sure that it will turn into a problem on Monday morning if it is not fixed before then. Although this distinction between symptoms and cause of a problem might seem philosophical, you need to be aware of the potential communication issues that might arise from it.

Generally, reporting of a problem triggers the troubleshooting process. Troubleshooting starts by defining the problem. The second step is diagnosing the problem, during which information is gathered, the problem definition is refined, and possible causes for the problem are proposed. Eventually, this process should lead to a hypothesis for the root cause of the problem. At this time, possible solutions need to be proposed and evaluated. Next, the best solution is selected and implemented. Figure 6.1 illustrates the main elements of a structured troubleshooting approach and the transition possibilities from one step to the next.

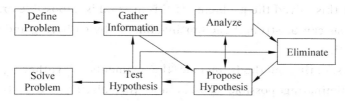

Figure 6.1　Flow Chart of a Structured Troubleshooting Approach.

NOTE

It is noteworthy, however, that the solution to a network problem cannot always be readily implemented and an interim workaround might have to be proposed. The difference between a solution and a workaround is that a solution resolves the root cause of the problem, whereas a workaround only alleviates the symptoms of the problem.

Although problem reporting and resolution are definitely essential elements of the troubleshooting process, most of the time is spent in the diagnostic phase. One might even believe that diagnosis is all troubleshooting is about. Nevertheless, within the context of network maintenance, problem reporting and resolution are indeed essential parts of troubleshooting. Diagnosis is the process of identifying the nature and cause of a problem. The main elements of this process are as follows:

- **Gathering information**: Gathering information happens after the problem has been reported by the user (or anyone). This might include interviewing all parties (user) involved, plus any other means to gather relevant information.

Usually, the problem report does not contain enough information to formulate a good hypothesis without first gathering more information. Information and symptoms can be gathered directly, by observing processes, or indirectly, by executing tests.

- **Analyzing information:** After the gathered information has been analyzed, the troubleshooter compares the symptoms against his knowledge of the system, processes, and baselines to separate normal behavior from abnormal behavior.
- **Eliminating possible causes:** By comparing the observed behavior against expected behavior, some of the possible problem causes are eliminated.
- **Formulating/proposing a hypothesis:** After gathering and analyzing information and eliminating the possible causes, one or more potential problem causes remain. The probability of each of these causes will have to be assessed and the most likely cause proposed as the hypothetical cause of the problem.
- **Testing the hypothesis:** The hypothesis must be tested to confirm or deny that it is the actual cause of the problem. The simplest way to do this is by proposing a solution based on this hypothesis, implementing that solution, and verifying whether this solved the problem. If this method is impossible or disruptive, the hypothesis can be strengthened or invalidated by gathering and analyzing more information.

All troubleshooting methods include the elements of gathering and analyzing information, eliminating possible causes, and formulating and testing hypotheses. Each of these steps has its merits and requires some time and effort; how and when one moves from one step to the next is a key factor in the success level of a troubleshooting exercise. In a scenario where you are troubleshooting a complex problem, you might go back and forth between different stages of troubleshooting: Gather some information, analyze the information, eliminate some of the possibilities, gather more information, analyze again, formulate a hypothesis, test it, reject it, eliminate some more possibilities, gather more information, and so on.

If you do not take a structured approach to troubleshooting and do troubleshooting in an ad hoc fashion, you might eventually find the solution; however, the process in general will be very inefficient. Another drawback of ad hoc troubleshooting is that handing the job over to someone else is very hard to do; the progress results are mainly lost. This can happen even if the troubleshooter wants to resume his own task after he has stopped for a while, perhaps to take care of another matter. A structured approach to troubleshooting, regardless of the exact method adopted, yields more predictable results in the long run. It also makes it easier to pick up where you left off or hand the job over to someone else without losing any effort or results.

A troubleshooting approach that is commonly deployed both by inexperienced and experienced troubleshooters is called shoot-from-the-hip. After a very short period of gathering information, taking this approach, the troubleshooter quickly makes a change to see if it solves the problem. Even though it may seem like random troubleshooting on the surface, it is not. The reason is that the guiding principle for this method is prior and usually vast knowledge of common symptoms and their corresponding causes, or simply extensive relevant experience in a particular environment or application. This technique might be quite effective for the experienced troubleshooter most times, but it usually does not yield the same results for the inexperienced troubleshooter. Figure 6.2 shows how the "shoot-from-the-hip" approach goes about solving a problem, spending almost no effort in analyzing the gathered information and eliminating possibilities.

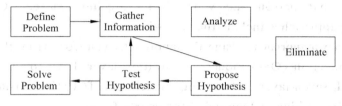

Figure 6.2 Shoot-from-the-Hip.

Assume that a user reports a LAN performance problem and in 90 percent of the past cases with similar symptoms, the problem has been caused by duplex mismatch between users' workstations (PC or laptop) and the corresponding access switch port. The solution has been to configure the switch port for 100-Mbps full duplex. Therefore, it sounds reasonable to quickly verify the duplex setting of the switch port to which the user connects and change it to 100-Mbps full duplex to see whether that fixes the problem. When it works, this method can be very effective because it takes very little time. Unfortunately, the downside of this method is that if it does not work, you have not come any closer to a possible solution, you have wasted some time (both yours and users'), and you might possibly have caused a bit of frustration. Experienced troubleshooters use this method to great effect. The key factor in using this method effectively is knowing when to stop and switch to a more methodical (structured) approach.

6.2 Structured Troubleshooting Approaches

Troubleshooting is not an exact science, and a particular problem can be diagnosed and sometimes even solved in many different ways. However, when you perform structured troubleshooting, you make continuous progress, and usually solve the problem faster than it would take using an ad hoc approach. There are many different structured troubleshooting approaches. For some problems, one method might work better, whereas for others, another method might be more suitable. Therefore, it is beneficial for the troubleshooter to be familiar with a variety of structured approaches and select the best method or combination of methods to solve a particular problem.

A structured troubleshooting method is used as a guideline through a troubleshooting process. The key to all structured troubleshooting methods is systematic elimination of hypothetical causes and narrowing down on the possible causes. By systematically eliminating possible problem causes, you can reduce the scope of the problem until you manage to isolate and solve the problem. If at some point you decide to seek help or hand the task over to someone else, your findings can be of help to that person and your efforts are not wasted. Commonly used troubleshooting approaches include the following:

- **The top-down approach**: Using this approach, you work from the Open Systems Interconnection (OSI) model's application layer down to the physical layer. The OSI seven-layer networking model and TCP/IP four-layer model are shown side by side in Figure 6.3 for your reference.

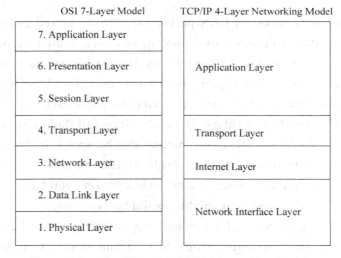

Figure 6.3 The OSI and TCP/IP Networking Models.

- **The bottom-up approach**: This approach starts from the OSI model's physical layer and moves up toward the application layer.
- **The divide-and-conquer approach**: Using this approach, you start in the middle of the OSI model's stack (usually the network layer), and then, based on your findings, you move up or down the OSI stack.
- **The follow-the-path approach**: This approach is based on the path that packets

take through the network from source to destination.
- **The spot-the-differences approach**: As the name implies, this approach compares network devices or processes that are operating correctly to devices or processes that are not operating as expected and gathers clues by spotting significant differences. In case the problem occurred after a change on a single device was implemented, the spot-the differences approach can pinpoint the problem cause by focusing on the difference between the device configurations, before and after the problem was reported.
- **The move-the-problem approach**: The strategy of this troubleshooting approach is to physically move components and observe whether the problem moves with the moved components.

The sections that follow describe each of these methods in more detail.

6.2.1 The Top-Down Troubleshooting Approach

The top-down troubleshooting method uses the OSI model as a guiding principle. One of the most important characteristics of the OSI model is that each layer depends on the underlying layers for its operation. This implies that if you find a layer to be operational, you can safely assume that all underlying layers are fully operational as well.

Let's assume that you are researching a problem of a user that cannot browse a particular website and you find that you can establish a TCP connection on port 80 from this host to the server and get a response from the server (see Figure 6.4). In this situation, it is reasonable to conclude that the transport layer and all layers below must be fully functional between the client and the server and that this is most likely a client or server problem (most likely at application, presentation, or session layer) and not a network problem. Be aware that in this example it is reasonable to conclude that Layers 1 through 4 must be fully operational, but it does not definitively prove this. For instance, nonfragmented packets might be routed correctly, whereas fragmented packets are dropped. The TCP connection to port 80 might not uncover such a problem.

Essentially, the goal of the top-down approach is to find the highest OSI layer that is still working. All devices and processes that work on that layer or layers below are then eliminated from the scope of the troubleshooting. It might be clear that this approach is most effective if the problem is on one of the higher OSI layers. It is also one of the most straightforward troubleshooting approaches, because problems reported by users are typically defined as application layer problems, so starting the troubleshooting process at that layer is a natural thing to do. A drawback or

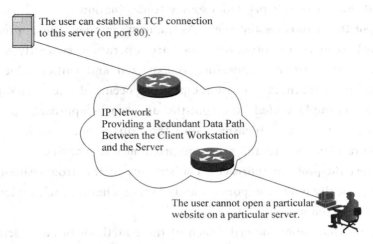

Figure 6.4   Application Layer Failure.

impediment to this approach is that you need to have access to the client's application layer software to initiate the troubleshooting process, and if the software is only installed on a small number of machines, your troubleshooting options might be limited.

6.2.2 The Bottom-Up Troubleshooting Approach

The bottom-up troubleshooting approach also uses the OSI model as its guiding principle with the physical layer (bottom layer of the OSI seven-layer network model) as the starting point. In this approach, you work your way layer by layer up toward the application layer and verify that relevant network elements are operating correctly. You try to eliminate more and more potential problem causes so that you can narrow down the scope of the potential problems.

Let's assume that you are researching a problem of a user that cannot browse a particular website and while you are verifying the problem, you find that the user's workstation is not even able to obtain an IP address through the DHCP process (see Figure 6.5). In this situation it is reasonable to suspect lower layers of the OSI model and take a bottom-up troubleshooting approach.

A benefit of the bottom-up approach is that all the initial troubleshooting takes place on the network, so access to clients, servers, or applications is not necessary until a very late stage in the troubleshooting process. In certain environments, especially those where many old and outdated devices and technologies are still in use, many network problems are hardware related. The bottom-up approach is very effective under those circumstances. A disadvantage of this method is that, in large

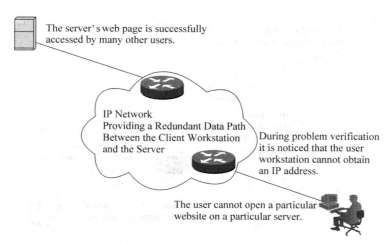

Figure 6.5　Failure at Lower OSI Layers.

networks, it can be a time-consuming process because a lot of effort will be spent on gathering and analyzing data and you always start from the bottom layer. The best bottom-up approach is to first reduce the scope of the problem using a different strategy and then switch to the bottom-up approach for clearly bounded parts of the network topology.

6.2.3　The Divide-and-Conquer Troubleshooting Approach

　　The divide-and-conquer troubleshooting approach strikes a balance between the top-down and bottom-up troubleshooting approaches. If it is not clear which of the top-down or bottom-up approaches will be more effective for a particular problem, an alternative is to start in the middle (usually from the network layer) and perform some tests such as ping and trace. Ping is an excellent connectivity testing tool. If the test is successful, you can assume that all lower layers are functional, and so you can start a bottom-up troubleshooting starting from the network layer. However, if the test fails, you can start a top-down troubleshooting starting from the network layer.

　　Let's assume that you are researching a problem of a user who cannot browse a particular website and that while you are verifying the problem you find that the user's workstation can successfully ping the server's IP address (see Figure 6.6). In this situation, it is reasonable to assume that the physical, data link, and network layers of the OSI model are in good working condition, and so you examine the upper layers, starting from the transport layer in a bottom-up approach.

　　Whether the result of the initial test is positive or negative, the divide-and-conquer approach usually results in a faster elimination of potential problems than what you would achieve by implementing a full top-down or bottom-up approach.

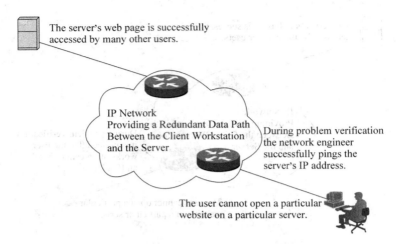

Figure 6.6 Successful Ping Shifts the Focus to Upper OSI Layers (Divide-and-Conquer Approach).

Therefore, the divide-and-conquer method is considered highly effective and possibly the most popular troubleshooting approach.

6.2.4 The Follow-the-Path Troubleshooting Approach

The follow-the-path approach is one of the most basic troubleshooting techniques, and it usually complements one of the other troubleshooting methods such as the top-down or the bottom-up approach. The follow-the-path approach first discovers the actual traffic path all the way from source to destination. Next, the scope of troubleshooting is reduced to just the links and devices that are actually in the forwarding path. The principle of this approach is to eliminate the links and devices that are irrelevant to the troubleshooting task at hand.

Let's assume that you are researching a problem of a user who cannot browse a particular website and that while you are verifying the problem you find that a trace (tracert) from the user's PC command prompt to the server's IP address succeeds only as far as the first hop, which is the L3 Switch v (Layer 3 or Multilayer Switch v) in Figure 6.7. Based on your understanding of the network link bandwidths and the routing protocol used on this network, you mark the links on the best path between the user workstation and the server on the diagram with numbers 1 through 7, as shown in Figure 6.7.

In this situation it is reasonable to shift your troubleshooting approach to the L3 Switch v and the segments beyond, toward the server along the best path. The follow-the-path approach can quickly lead you to the problem area. You can then try and pinpoint the problem to a device, and ultimately to a particular physical or logical

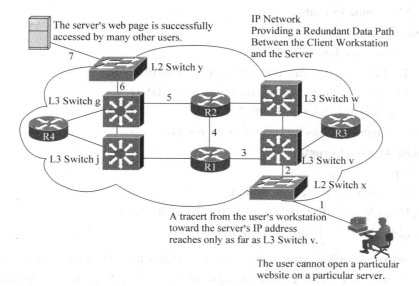

Figure 6.7　The Follow-the-Path Approach Shifts the Focus to Link 3 and Beyond Toward the Server.

component that is either broken, misconfigured, or has a bug.

6.2.5　The Compare-Configurations Troubleshooting Approach

　　Another common troubleshooting approach is called the compare-configurations approach, also referred to as the spotting-the-differences approach. By comparing configurations, software versions, hardware, or other device properties between working and nonworking situations and spotting significant differences between them, this approach attempts to resolve the problem by changing the nonoperational elements to be consistent with the working ones. The weakness of this method is that it might lead to a working situation, without clearly revealing the root cause of the problem. In some cases, you are not sure whether you have implemented a solution or a workaround.

　　Example 6.1 shows two routing tables; one belongs to Branch2's edge router, experiencing problems, and the other belongs to Branch1's edge router, with no problems. If you compare the content of these routing tables, as per the compare-configurations (spotting-the-differences) approach, a natural deduction is that the branch with problems is missing a static entry. The static entry can be added to see whether it solves the problem.

　　Example 6.1　*Spot-the-Differences*: *One Malfunctioning and One Working Router*
　　------------- Branch1 is in good working order ----------

```
Branch1# show ip route
<...output omitted...>
10.0.0.0/24 is subnetted, 1 subnets
C      10.132.125.0 is directly connected, FastEthernet4
C      192.168.36.0/24 is directly connected, BVI1
S*     0.0.0.0/0 [254/0] via 10.132.125.1
------------ Branch2 has connectivity problems ----------
Branch2# show ip route
<...output omitted...>
10.0.0.0/24 is subnetted, 1 subnets
C 10.132.126.0 is directly connected, FastEthernet4
C 192.168.37.0/24 is directly connected, BVI1
```

The compare-configurations approach (spotting-the-differences) is not a complete approach; it is, however, a good technique to use undertaking other approaches. One benefit of this approach is that it can easily be used by less-experienced troubleshooting staff to at least shed more light on the case. When you have an up-to-date and accessible set of baseline configurations, diagrams, and so on, spotting the difference between the current configuration and the baseline might help you solve the problem faster than any other approach.

6.2.6 The Swap-Components Troubleshooting Approach

Also called move-the-problem, the swap-components approach is a very elementary troubleshooting technique that you can use for problem isolation: You physically swap components and observe whether the problem stays in place, moves with the component, or disappears entirely. Figure 6.8 shows two PCs and three laptops connected to a LAN switch, among which laptop B has connectivity problems. Assuming that hardware failure is suspected, you must discover whether the problem is on the switch, the cable, or the laptop. One approach is to start gathering data by checking the settings on the laptop with problems, examining the settings on the switch, comparing the settings of all the laptops, and the switch ports, and so on. However, you might not have the required administrative passwords for the PCs, laptops, and the switch. The only data that you can gather is the status of the link LEDs on the switch and the laptops and PCs. What you can do is obviously limited. A common way to at least isolate the problem (if it is not solved outright) is cable or port swapping. Swap the cable between a working device and laptop B (the one that is having problems). Move the laptop from one port to another using a cable that you know for sure is good. Based on these simple moves, you can isolate whether the problem is cable, switch, or laptop related.

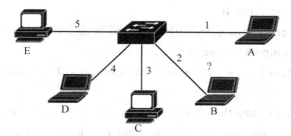

Figure 6.8　Swap-the-Component: Laptop B Is Having Network Problems.

Just by executing simple tests in a methodical way, the swap-components approach enables you to isolate the problem even if the information that you can gather is minimal. Even if you do not solve the problem, you have scoped it to a single element, and you can now focus further troubleshooting on that element. Note that in the previous example if you determine that the problem is cable related, it is unnecessary to obtain the administrative password for the switch, PCs, and laptops. The drawbacks of this method are that you are isolating the problem to only a limited set of physical elements and not gaining any real insight into what is happening, because you are gathering only very limited indirect information. This method assumes that the problem is with a single component. If the problem lies within multiple devices, you might not be able to isolate the problem correctly.

6.3　Troubleshooting Example Using Six Different Approaches

An external financial consultant has come in to help your company's controller with an accounting problem. He needs access to the finance server. An account has been created for him on the server, and the client software has been installed on the consultant's laptop. You happen to walk past the controller's office and are called in and told that the consultant can't connect to the finance server. You are a network support engineer and have access to all network devices, but not to the servers. Think about how you would handle this problem, what your troubleshooting plan would be, and which method or combination of methods you would use.

What possible approaches can you take for this troubleshooting task? This case lends itself to many different approaches, but some specific characteristics can help you decide an appropriate approach:
- You have access to the network devices, but not to the server. This implies that you will likely be able to handle Layer 1-4 problems by yourself; however, for Layer 5-7, you will probably have to escalate to a different person.

- You have access to the client device, so it is possible to start your troubleshooting from it.
- The controller has the same software and access rights on his machine, so it is possible to compare between the two devices.

What are the benefits and drawbacks of each possible troubleshooting approach for this case?

- **Top-down**: You have the opportunity to start testing at the application layer. It is good troubleshooting practice to confirm the reported problem, so starting from the application layer is an obvious choice. The only possible drawback is that you will not discover simple problems, such as the cable being plugged in to a wrong outlet, until later in the process.
- **Bottom-up**: A full bottom-up check of the whole network is not a very useful approach because it will take too much time and at this point, there is no reason to assume that the network beyond the first access switch would be causing the issue. You could consider starting with a bottom-up approach for the first stretch of the network, from the consultant's laptop to the access switch, to uncover potential cabling problems.
- **Divide-and-conquer**: This is a viable approach. You can ping from the consultant's laptop to the finance server. If that succeeds, the problem is most likely at upper layers. For example, a firewall or access control list could be the culprit. If the ping fails, assuming that ping is not blocked in the network, it is safe to assume that the problem is at network or lower layers and you are responsible for fixing it. The advantage of this method is that you can quickly decide on the scope of the problem and whether escalation is necessary.
- **Follow-the-path**: Similar to the bottom-up approach, a full follow-the-path approach is not efficient under the circumstances, but tracing the cabling to the first switch can be a good start if it turns out that the link LED is off on the consultant's PC. This method might come into play after other techniques have been used to narrow the scope of the problem.
- **Compare-configurations**: You have access to both the controller's PC and the consultant's laptop; therefore, compare-configurations is a possible strategy. However, because these machines are not under the control of a single IT department, you might find many differences, and it might therefore be hard to spot the significant and relevant differences. The compare-configurations approach might prove useful later, after it has been determined that the problem is likely to be on the client.
- **Swap-components**: Using this approach alone is not likely to be enough to solve the problem, but if following any of the other methods indicates a potential

hardware issue between the consultant's PC and the access switch, this method might come into play. However, merely as a first step, you could consider swapping the cable and the jack connected to the consultant's laptop and the controller's PC, in turn, to see whether the problem is cable, PC, or switch related.

Many combinations of these different methods could be considered here. The most promising methods are top-down or divide-and-conquer. You will possibly switch to follow-the-path or compare-configurations approach after the scope of the problem has been properly reduced. As an initial step in any approach, the swap-components method could be used to quickly separate client-related issues from network-related issues. The bottom-up approach could be used as the first step to verify the first stretch of cabling.

6.4 Summary

The fundamental elements of a troubleshooting process are as follows:
- Defining the problem
- Gathering information
- Analyzing information
- Eliminating possible causes
- Formulating a hypothesis
- Testing the hypothesis
- Solving the problem

Some commonly used troubleshooting approaches are as follows:
- Top-down
- Bottom-up
- Divide-and-conquer
- Follow-the-path
- Compare-configurations
- Swap-components

6.5 New Words and Expressions

daunting	adj. 使人畏缩的;令人却步的
sophisticated	adj. 复杂的;精致的
turn into	v. 变成;进入
hypothesis	n. 假设
interim	adj. 临时的,暂时的;中间的

alleviate	*vt.* 减轻,缓和
proposing a hypothesis	提出一个假设
scenario	*n.* 方案;情节
back and forth	反复地,来回地
ad hoc fashion	特定方式
full duplex	全双工
OSI(open systems interconnection)	开放系统互连
divide-and-conquer approach	"分而治之"法
alternative	*n.* 二中择一;供替代的选择
routing protocol	路由协议

6.6 Questions

Single Choice Questions

1. Which of the following statements is true? (　　)
   A. The more efficiently and effectively the network support personnel diagnose and resolve problems, the lower impact and damages will be to business
   B. The more efficiently and effectively the network support personnel diagnose and resolve problems, the higher impact and damages will be to business
   C. The lower impact and damages will be to business, the more efficiently and effectively the network support personnel diagnose and resolve problems
   D. The higher impact and damages will be to business, the more efficiently and effectively the network support personnel diagnose and resolve problems

2. What does troubleshooting starts by? (　　)
   A. Analyzing the problem
   B. Defining the problem
   C. Proposing the problem
   D. Solving the problem

3. What is done before one or more potential problem causes remain? (　　)
   A. Gathering and analyzing information
   B. Gathering information and eliminating the possible causes
   C. Gathering and analyzing information and eliminating the possible causes
   D. None of the above

4. Figure 6.2 shows how the "shoot-from-the-hip" approach goes about except which of the following? (　　)

A. Solving a problem

B. Spending almost no effort in analyzing the gathered information

C. Eliminating possibilities

D. Reporting of a problem

5. Which method is considered highly effective and possibly the most popular troubleshooting approach? (   )

  A. The divide-and-conquer method

  B. The top-down troubleshooting method

  C. The bottom-up troubleshooting approach

  D. All of the above

6. Which of the following statements is true? (   )

  A. The top-down approach is one of the most basic troubleshooting techniques, and it usually complements one of the other troubleshooting methods such as the follow-the-path or the bottom-up approach

  B. The follow-the-path approach is one of the most basic troubleshooting techniques, and it usually complements one of the other troubleshooting methods such as the top-down or the bottom-up approach

  C. The bottom-up approach is one of the most basic troubleshooting techniques, and it usually complements one of the other troubleshooting methods such as the top-down or the follow-the-path approach

  D. The divide-and-conquer approach is one of the most basic troubleshooting techniques, and it usually complements one of the other troubleshooting methods such as the top-down or the bottom-up approach

7. Which of the following processes is not the subprocesses or phase of a troubleshooting process? (   )

  A. Solve the problem   B. Eliminate

  C. Compile   D. Define the problem

8. Which of the following approaches is invalid troubleshooting methods? (   )

  A. Swap-components   B. Compare-configurations

  C. Follow-the-path   D. Hierarchical

9. Which of the following troubleshooting approaches doesn't use the OSI reference model as a guiding principle? (   )

  A. Top-down   B. Bottom-up

  C. Divide-and-conquer   D. Swap-components

10. Which of the following troubleshooting methods would be most effective when the problem is with the Ethernet cable connecting a workstation to the wall RJ-45 jack? (   )

  A. Divide-and-conquer   B. Compare-configurations

C. Swap-components   D. Follow-the-path

6.7 Problems

After reading this chapter and completing the exercises, you will be able to do the following:

1. Describe troubleshooting principles.
2. Describe common troubleshooting approaches.
3. Discuss the main elements of this process.
4. Explain structured troubleshooting approaches.
5. Describe each of these methods in detail.
6. What are the benefits and drawbacks of each possible troubleshooting approach for this case?

Unit 7 Mobile Application—What's Special about Mobile Testing?

This chapter explains that mobile testing is completely different from testing on other technologies such as laptops or desktop computers. The mobile user is on the move while he or she is using your product. Therefore, it is very important to know about the different data networks and the different types of mobile devices. This chapter also provided a first overview of mobile users' high expectations. And last but not least, this chapter should remind you to never underestimate a new technology. Be open-minded and curious to improve your daily work life.

Before I start describing the unique aspects of mobile testing, I'd like to share a true story with you.

What's special about mobile testing? Someone asked me this exact question several years ago while at a testing conference. I started talking about mobile technologies, apps, how to test them, and what's special about mobile testing. The guy simply smiled at me and said, "But it's software just on a smaller screen. There's nothing special about it." He was really arrogant and didn't see the challenges presented by mobile testing. No matter which arguments I used to convince him, he didn't believe in the importance of mobile technologies, apps, and testing.

I met the same guy again in 2014 while at a testing conference where he talked about mobile testing. He spoke about the importance of apps and how important it is to test them.

As you can see, it's very easy to underestimate new technologies. As a software tester it's especially helpful to be curious about learning something new and exploring new technologies to broaden your skills.

So let's come back to the initial question: What's special about mobile testing? I think I can assume you have at least one mobile device, namely, a smartphone. Or maybe you have a tablet, or even both. If you look at your device(s), what do you see? Just a small computer with little shiny icons on its screen? Or do you see a very personal computer with lots of sensors and input options that contains all of your private data? Please take a minute to think about that.

My smartphone and tablet are very personal computers that hold almost all of my data, be it e-mails, SMS, photos, music, videos, and the like. I can access my data no matter where I am and use my smartphone as a navigation and information system to find out more about my surroundings. For that reason I expect my apps to be reliable, fast, and easy to use.

In those three sentences I described my personal expectations of mobile devices and apps. But you may have entirely different expectations, as does the next person. And this brings me to the first special characteristic or unique aspect of mobile testing: user expectations.

7.1 User Expectations

In my opinion, the user of an app is the main focus and main challenge for mobile teams. The fact that every user has unique expectations makes it difficult to develop and deliver the "right" app to customers. As several reports and surveys have shown, mobile users have far higher expectations of mobile apps than of other software such as browser applications. The majority of reports and surveys state that nearly 80% of users delete an app after using it for the first time! The top four reasons for deletion are always bad design, poor usability, slow loading time, and crashes immediately after installation. Nearly 60% of users will delete an app that requires registration, and more than half of users expect an app to launch in under two seconds. If the app takes more time, it gets deleted. Again, more than half of users experience crashes the very first time they start an app. An average user checks his or her mobile device every six minutes and has around 40 apps installed. Based on those numbers, you can deduce that mobile users have really high expectations when it comes to usability, performance, and reliability. Those three characteristics were mentioned most often by far when users were asked about their experience with mobile apps.

Currently there are more than two million apps available in the app stores of the biggest vendors. A lot of apps perform the same task, meaning that there's always at least one competitor app, which makes it very easy for consumers to download a different app as it's just a single tap away. Here are some points you should keep in mind when developing and testing a mobile app:

- Gather information about your possible target customer group.
- Ask your customers about their needs.
- Your app needs to solve a problem for the user.
- Usability is really important.
- Your app needs to be reliable and robust.
- App performance is really important.
- Apps need to be beautiful.

There are, of course, a plethora of other things you should take into account, but if you pay attention to these points, your users are likely to be happy.

You've probably already heard of the Expectations principle. KISS is an acronym for Keep It Simple, Stupid and is always a useful reminder—especially for software projects—to not inflate the software with just another function or option. Keeping it small, easy, and simple is best in most cases and is likely to make your customers happy. Inspired by KISS, I came up with my own principle for mobile apps: KIFSU (see Figure 7.1). This abbreviation is a good mnemonic to help you cover customer needs and a constant reminder not to inflate apps with useless functions.

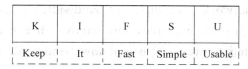

Figure 7.1 KIFSU.

7.2 Mobility and Data Networks

Another challenge mobile apps have to deal with more than software running on computers is the fact that users are moving around while they use apps, which often requires an Internet connection to fetch data from the backend and serve the user with updates and information.

Mobile apps need to be tested in real life, in real environments where the potential user will use them. For example, if you're testing an app for snowboarders and skiers that accesses slope information, one that is able to record the speed of the current downhill run and makes it possible for users to share records directly with their friends, you need to test these functions on a slope. Otherwise you can't guarantee that every feature will work as expected.

Of course, there are parts of an app that you can test in a lab situation, such as slope information availability or whether or not the app can be installed, but what about recording a person's speed, the weather conditions, or the Internet connection at the top of a mountain?

The weather conditions on a mountain, in particular, can be very difficult to handle as they can, of course, range from sunshine to a snowstorm. In such scenarios you will probably find lots of bugs regarding the usability and design of an app. Maybe you'll also find some functional bugs due to the temperature, which may have an impact on your hardware and, in turn, your app.

As I already mentioned, the speed and availability of Internet connections could vary in such regions. You will probably have a good network connection with high speed at the top of the mountain and a really poor one down in the valley. What happens if you have a bad or no Internet connection while using the app? Will it crash or will it still work? What happens if the mobile device changes network providers while the app is being used? (This is a common scenario when using apps close to an international border, such as when snowboarding in the Alps.)

All of these questions are very hard to answer when testing an app in a lab. You as a mobile tester need to be mobile and connected to data networks while testing apps.

As you can see, it's important to test your app in real-life environments and to carry out tests in data networks with different bandwidths as the bandwidth can have a huge impact on your app; for example, low bandwidth can cause unexpected error messages, and the switch between high and low bandwidth can cause performance issues or freezes.

Here's an exercise for you. Take any app you want and find three usage scenarios where the environment and/or network connection could cause problems.

7.3 Mobile Devices

Before you continue reading, pick up your mobile device and look at it. Take your device in your hand and look at every side of it without turning it on. What do you see?

You will most likely see a device with a touch-sensitive screen, a device with several hardware buttons with a charger, a headphone connection, and a camera. That's probably it—you're not likely to have more than five hardware buttons (except for smartphones with a physical keyboard).

In an era when the words *cell phone* have become synonymous with smartphone, it's important to remember that there used to be other types of cell phones, so-called dumb phones and feature phones that have lots more hardware buttons for making a call or typing a message. With a conventional dumb phone you are only able to make a call, type a message, or store a contact list; they're not usually connected to the Internet. The more advanced ones, the feature phones, have games, a calendar, or a very basic Web browser with the option to connect to the Internet. But all these phones are really basic in terms of functionality and expandability as users aren't able to install apps or easily update the software to a newer version, if it all. Both types of phones are still available, especially in emerging markets, but since 2013 more smartphones have been sold worldwide than dumb phones or feature phones, and this trend is likely to continue as time goes on. In fact, in the next couple of years dumb phones and feature phones will be a thing of the past.

The phones we use nowadays are completely different from the "old" ones. Current smartphones are mini supercomputers with lots of functionality in terms of hardware and software. They're packed with various sensors such as brightness, proximity, acceleration, tilt, and much more. Besides that, all modern smartphones have both front- and rear-facing cameras, various communication interfaces such as Bluetooth, near field communication (NFC), and Global Positioning System (GPS), as well as Wi-Fi and cellular networks to connect to the Internet. Depending on the mobile platform and mobile manufacturer, you may find an array of other hardware features.

From a software point of view, smartphones offer lots of application programming interfaces (APIs) for manufacturers, developers, and users to extend smartphone capabilities with apps.

If you just focus on the major mobile platforms, iOS and Android, there are plenty of hardware and software combinations that mobile testers have to deal with. The fact that there are so many combinations is known as fragmentation. Mobile device fragmentation is a huge topic and yet another challenge when it comes to mobile testing.

You can't test your app with every possible hardware and software combination. And the fact that you should test your app in a real environment makes it even more impossible. Mobile testers need to find a strategy to downsize the effort of testing on different devices and to find a way to test on the right devices.

But how can that be accomplished? By testing on just one mobile platform? By testing on just the latest device? By testing with just the latest software version?

Before you define a strategy, you should keep in mind that every app is unique, has unique requirements, has other problems to solve, and has a unique user base. With these points in mind, you can ask yourself the following questions to find the "right" mobile devices for testing:

- Who is my user base?
- How old is the average user?
- How many men or women are in my target user group?
- Which platform is used most among that user base?
- Which device is used most?
- Which software version is installed on most of the phones?
- What kind of sensors does my app use?
- How does the app communicate with the outside world?
- What is my app's main use case?

Of course, there are lots more questions to ask, but if you answer most of the ones I suggest, the list of possible devices you should consider testing is much shorter.

7.4　Mobile Release Cycles

Now that you know how to find the right devices for testing your app, it doesn't mean that the process is over. To be honest, it's never going to end!

The main mobile manufacturers release a new flagship phone with more features every year. In and around those releases they bring out other phones for different user scenarios and user groups. This is especially true in the Android world where every new phone comes with a new version of the operating system packed with new features, designs, or APIs. There are multiple software releases within the course of a year, ranging from bug fixes to feature releases. You as a mobile tester need to be sure that your app will run on the latest hardware and software.

But how should you handle these situations? By buying every phone that appears on the market? By constantly updating to the latest operating system version?

Again, the most important factors are your target customer group and the app you're testing. When you know that your target group always uses the latest and fastest phones on the market, you need to buy those phones as soon as they appear. Regardless of whether or not your target group is up-to-date, you should always monitor the mobile market.

You need to know when the main vendors are due to release new flagship phones that a lot of people are likely to buy. You also need to know when the operating systems receive patches, new features, or new design patterns.

So the answer to the question of whether you need to buy every phone and constantly update the operating systems is yes and no. Of course you don't need to buy every phone that's on the market, but you should consider updating to the latest operating system version. When doing so, keep in mind that not every user will install the update. Many people don't know how to do that, or they don't care about new versions. You need at least some phones that are running older versions of the operating system to see how the app reacts in that environment. Older versions of the operating system are also needed to reproduce reported problems and bugs.

A good way to manage all this is to stick with the same operating system version on the phones that you have and buy new phones with the latest software version. This of course leads to another problem—it's really expensive! Not every manager wants to spend so much money on mobile devices when a phone is going to be used for only a couple of months. A solution for that is to rent devices. There are several providers and Open Device Labs where you can rent a device for a certain period of time. Another way to rent devices is the mobile device cloud as there are a number of providers who give mobile testers exclusive access to the physical devices they have made available in the cloud. Just use your search engine and check them out.

In the mobile projects I've worked on, we always had the top ten to 15 devices used by our target user group in different variations for developing and testing. This was a good number of devices that covered nearly 90% of our target group. With those ten to 15 devices we were able to find most of the critical bugs; the remaining 10% of devices we didn't have were of no major consequence to the project or user expectations.

In order to handle the fast pace of mobile release cycles, you should keep the following things in mind:
- Monitor the mobile device and software market.
- Know when new phones will be rolled out.
- Find out about the new features of the operating systems.
- Keep an eye on your target customer group to see if new devices are showing up in your statistics.
- Think twice before updating a phone to the latest operating system version.
- Buy new phones with the latest operating system version.
- If buying is not an option, rent the devices.

Updating, buying, and maintaining all of your devices is a challenging task and should not be underestimated! At some point, depending on the number of test devices used within a project, this could be a full-time job.

7.5　Mobile Testing Is Software Testing

Let's come back to the story I told at the beginning of this chapter when the guy at the conference didn't believe in the importance of mobile testing. He had the attitude that mobile testing is not real software testing. In his opinion, mobile apps were only small programs with less functionality and no real challenges when it comes to software testing. But this is definitely not the case. If you look at the topics I described in this chapter, you should have an initial impression about the challenging job of a mobile tester. Mobile testing is totally different from testing software applications such as Web or desktop applications. With mobile apps, physical devices have far more influence over the software that is running on them when compared to other software such as Web applications. Because there are so many different smartphones available on the market, mobile testers need to focus a lot more on hardware during the testing process. In addition, users moving around and using different data networks force mobile testers to be on the move while testing.

Besides the hardware, user expectations play an important part in the daily business of a mobile tester and need to be taken seriously.

There are many more topics and issues mobile testers need to know about in order to help the whole team release a successful app. The rest of the chapters in this book will cover the following topics:

- More challenges for mobile testers and solutions to those challenges
- How to test mobile apps systematically
- How to select the right mobile test automation tool
- The different concepts of mobile test automation tools
- How to find the right mobile testing strategy
- Additional mobile testing methods
- Required skills for mobile testers

Keep the topics from this chapter in mind as a starting point. Keep your app simple and fast (remember KIFSU). Test while you're on the move, and test on different devices based on your target customer group.

7.6 Summary

As you have seen, mobile testing is completely different from testing on other technologies such as laptops or desktop computers. The biggest difference between mobile and other technologies is that the mobile user is on the move while he or she is using your product. Therefore, it is very important to know about the different data networks and the different types of mobile devices.

This chapter also provided a first overview of mobile users' high expectations. It is really important to keep KIFSU in mind when designing, developing, and testing a mobile app. It will help you to focus on the important elements and not waste time on unnecessary features that your users won't use.

And last but not least, this chapter should remind you to never underestimate a new technology. Be open-minded and curious to improve your daily work life.

7.7 New Words and Expressions

be curious about	对…感到好奇
registration	n. 登记；注册
mnemonic	n. 记忆术；助记符
guarantee	vt. 保证；担保
have an impact on	对于…有影响；对…造成冲击
NFC(Near Field Communication)	近场通信
stick with	坚持；继续做；保持联系

consequence	*n.* 结果；推论
be rolled out	被推出
underestimate	*vt.* 低估；看轻
take seriously	认真对待

7.8 Questions

Single Choice Questions

1. How many percent of users is it who delete an app after using it for the first time stated by the majority of reports and surveys? (　　)

    A. 60%　　　　B. 70%　　　　C. 80%　　　　D. 90%

2. What always are the top four reasons for deletion? (　　)

    A. Bad design, poor usability, slow loading time, and crashes immediately after installation

    B. High price, poor usability, slow loading time, and crashes immediately after installation

    C. Bad design, high price, slow loading time, and crashes immediately after installation

    D. Bad design, poor usability, high price, and crashes immediately after installation

3. Which of the following statements is true? (　　)

    A. Based on those numbers, you can deduce that mobile users have really high expectations when it comes to usability

    B. Based on those numbers, you can deduce that mobile users have really high expectations when it comes to performance

    C. Based on those numbers, you can deduce that mobile users have really high expectations when it comes to reliability

    D. Based on those numbers, you can deduce that mobile users have really high expectations when it comes to usability, performance, and reliability

4. Which three characteristics were mentioned most often by far when users were asked about their experience with mobile apps? (　　)

    A. Security, performance, and reliability

    B. Usability, performance, and reliability

    C. Usability, security, and reliability

    D. Usability, performance, and security

5. In the KISS principle, what is KISS for as an acronym? (　　)

    A. Keep It Simple, Stupid

B. Keep It Simple, Strong

C. Keep It Simple, Smart

D. Keep It Simple, Slow

6. Which of the following statements is true? (   )

   A. Current smartphones are packed with various sensors such as brightness

   B. Current smartphones are packed with various sensors such as brightness and proximity

   C. Current smartphones are packed with various sensors such as brightness, proximity and acceleration

   D. Current smartphones are packed with various sensors such as brightness, proximity, acceleration, tilt, and much more

7. Which of the following statements is true? (   )

   A. The main mobile manufacturers release a new flagship phone with more features every year

   B. The main mobile manufacturers release a new flagship phone with more features every day

   C. The main mobile manufacturers release a new flagship phone with more features every week

   D. The main mobile manufacturers release a new flagship phone with more features every month

8. When you know that your target group always uses the latest and fastest phones on the market, what do you need to do as soon as they appear? (   )

   A. Buy those phones

   B. Sell those phones

   C. Use the phones

   D. None of the above

9. What are older versions of the operating system also needed to do? (   )

   A. Reproduce reported problems and bugs

   B. Reproduce reported features

   C. Reproduce reported designs

   D. Reproduce reported APIs

10. In order to handle the fast pace of mobile release cycles, which of the following things do you need not to keep in mind? (   )

    A. Monitor the mobile device and software market

    B. Know when new phones will be rolled out

    C. Find out about the features of the old operating systems

    D. If buying is not an option, rent the devices

7.9 Problems

After reading this chapter and completing the exercises, you will be able to do the following:

1. What's Special about Mobile Testing?
2. Discuss User Expectations.
3. Distinguish between Mobility and Data Networks.
4. Explain mobile devices.
5. Discuss Mobile Release Cycles.
6. Distinguish between mobile testing and soft testing.

Unit 8 Web Development—The Mobile Commerce Revolution and the Current State of Mobile

The statistics underline what is truly powerful about mobile technology: It's not about enabling things we couldn't have imagined. Instead, today's mobile technology enables exactly what the average consumer could have imagined, albeit in settings and situations that could not have been predicted.

Smartphones have also effectively ended the bar bet. Remember Norm and Cliff from the television series *Cheers*? Cliff's character, a mailman played by John Ratzenberger, was often the instigator and arbiter of trivia questions at the show's eponymous bar. Today, however, we are *all* Cliff. The answer to nearly any trivia question is just a mobile search away. In fact, having the answers in our pocket to "who sang that song?" and "who was that actor?" has conditioned us to expect the answers to nearly everything, whenever and wherever we need them. And that has changed our behavior, as we discuss later in the book.

8.1 Americans and Smartphones

Tom's company, Edison Research, has been tracking mobile phone ownership, usage, and other mobile behaviors for nearly a decade in its annual Infinite Dial study, a long-running research series that has been providing representative data about technology and media usage in America since 1998. In the most recent Infinite Dial study from 2014, an estimated 160 million Americans, or 61% of Americans aged 12 and older, now own a smartphone (defined as an Android, iOS, Blackberry, or Windows-based phone). What this number masks, however, are some significant demographic and psychographic differentials in smartphone ownership. In fact, the numbers for smartphone ownership for people aged 12 to 34 are truly staggering—more than three-quarters in that demographic (and 74% of teens!) own a smartphone.

What's truly interesting about smartphones (as compared to feature phones, which are non-smart phones without Internet access) is that a new generation of users has become as comfortable communicating with their thumbs as with their voices. When we asked mobile phone users how often they send or receive text messages on their phones, 75% of smartphone owners said several times per day, compared to just

29% of feature phone owners. Is this because smartphone owners are more communicative? Possibly, but unlikely. It's far more likely the power of a smartphone and the increased usage of apps on the mobile Internet makes users increasingly comfortable keeping the phone in front of them, rather than glued to their ear.

Indeed, we now have many means of communication available that simply didn't exist five years ago. Consider apps like Instagram and Snapchat. Three in 10 smartphone owners have Instagram accounts, making the company's $1 billion sale to Facebook look like a bargain in retrospect, especially when you consider that Facebook paid $19 billion for the messaging app WhatsApp. Snapchat, which has only been around for three years as of this writing, is now used by 19% of smart-phone users. In fact, as shown in Figure 8.1, nearly half of 12- to 24-year-olds use Snapchat (so the odds are good that your teenager does, too). Both services now tap into tens of millions of users who are using their phones to share images, messages, and above all, experiences in altogether new ways.

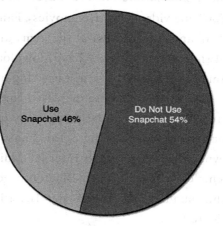

Figure 8.1 Snapchat usage amongst Americans 12-24.

Nowhere is this more apparent than in the use of social media. An incredible 40% of social media users with smartphones check those sites and services *several times per day*. This translates to some fairly remarkable behaviors. For example, the average smartphone owner who has a profile on Facebook checks his account six times per day, and 60% of them say they access Facebook *most* via mobile phone.

Previously, we noted that 83% of smartphone owners have their phones nearly always within arm's reach. This has resulted in many tens of millions relying on the phone as the first thing they look at in the morning (ostensibly after their spouse, as applicable)—indeed, for many smartphone owners, it is their device that actually wakes them up, replacing the clock radio. In fact, when asked what media they typically consume *most* at home in the morning, 24% of smartphone users indicated that it is their mobile device, second only to their television at 27%.

Of course, these stats can be slightly deceptive—after all, the smartphone has become the television, the radio, and the newspaper for so many. For the first time since Edison Research started tracking this stat in 2005, iPod ownership has actually declined, from 31% in 2013 to 29% in 2014. Today's smartphone owners have started to essentially replace their dedicated music players with their phones, and as a result, their media consumption habits have also changed.

For example, have you ever listened to a podcast? Before smartphones became so ubiquitous, mobile consumption of a podcast consisted of downloading a media file to your desktop or laptop, and then transferring it over to a portable media player to listen to it on the go. Today, all the friction has gone out of this process. In 2014, for the first time, most podcast users report that they listen primarily to podcasts on their mobile devices, and not on a computer.

Smartphones have opened up media consumption opportunities for audio, video, and text that heretofore never existed (or were at least difficult propositions). YouTube videos, Netflix movies, Pandora radio stations—all are available at the touch of an app, on the bus, at the gym, and even at the bar, next to Cliff. In fact, 50% of all smartphone owners have downloaded the Pandora app, an estimated 80 million Americans aged 12 + —a staggering number for an individual brand.

Nowhere are those increased consumption opportunities more apparent than in the car. The connected dashboard may or may not be a feature in your next car, but the fact is—it's already here for smartphone owners. In fact, 26% of mobile phone owners have listened to Internet radio in the car by connecting their phones to their vehicles (either through Bluetooth, or simply through a cable in an auxiliary jack). This number has grown significantly over the past few years, as you can see in Figure 8.2.

Online Radio Listening in a Car Via
Cell Phone Continues Steady Increase to 26%
% of Cell Phone Owners Who Have Ever Listened to Online Radio in a Car by
Listening to the Stream from a Cell Phone Connected to a Car Audio System

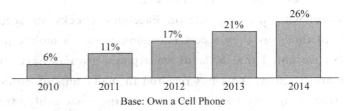

Base: Own a Cell Phone

Online Radio=Listening to AM/FM radio stations online and/or listening to streamed radio content available only on the Internet

Figure 8.2　In-car usage of Internet Radio.

This has opened up a world of opportunity for media producers to have their content consumed in new places, in new settings, and by multiple people (by freeing their content from the confines of the earbud). As a result, overall media consumption continues to rise; we are consuming media nearly every waking moment.

8.2 Mobile Around the World

The mobile revolution is a truly global revolution; indeed, Cisco's 2014 Global Mobile Data Traffic Forecast shows that global mobile data usage increased by 81% year-over-year from 2012 to 2013. Indeed, while North America continues to see significant growth, Cisco reports even higher growth percentages in Asia, and continuing growth throughout the developing world, as Figure 8.3 illustrates.

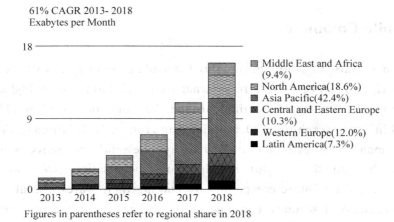

Figure 8.3 The compound annual growth rate (CAGR) in mobile data usage listed in exabytes (one exabyte = one billion gigabytes). Note the significant CAGR in the Asia Pacific region.

The really intriguing statistic in this report, however, lies in the distribution of mobile data. In 2013, the top 1% of mobile data users globally generated 10% of mobile data traffic. This is actually down 50% since 2010. What does this mean? Simply put, the evening out of mobile data usage worldwide indicates that more and more people are relying on their smartphones for everyday Internet-related tasks, reducing the relative contribution of the most active users.

The International Telecommunication Union (ITU) recently published some statistics on global mobile technology penetration. According to its most recent report, there are now 6.8 billion mobile-cellular subscriptions—almost one for every human on the planet. In fact, in developed nations, there is more than one mobile phone per resident (128%), with developing nations not far behind (89%).

Much has been said about the "platform wars" between Google's Android, Apple's iOS, Microsoft's Windows Mobile, and BlackBerry OS, and the statistics here vary considerably. While "usage" shows iOS as the leading mobile operating system, sales figures show Android-based phones outselling the field by a considerable margin in 2013. What is more important, however, than simply looking at phones as iOS devices, or Android devices, is this: They are all Internet devices. According to the Pew Internet and American Life research series, 34% of mobile Internet users admit that their mobile phone is their primary Internet access device, 4 a trend that is accelerating even faster globally. So the intense competition between mobile operating systems is truly having one positive benefit for the world: As these devices become more and more powerful and easy to use, they are truly putting the Internet into the world's pockets.

8.3 Mobile Commerce

Mobile penetration has grown rapidly, but mobile commerce is where some truly eye-popping figures can be found. In Internet Retailer's 2014 Mobile 500 study, 2013 mobile sales for the 500 largest retailers around the world increased by 71% over the same period in 2012, reaching $30.5 billion (up from $17.8 billion). Consider this stat for a moment. A growth rate of 71% for mobile retail commerce is significantly higher than the growth rate for mobile phone penetration. We can draw two conclusions here: First, more companies are making their wares available to sell via mobile channels. And second, consumers are becoming more comfortable making mobile purchases. Indeed, for many retailers, mobile is not only their biggest growth engine, it's becoming their biggest segment of sales, period.

In fact, we can already see mobile commerce behaviors becoming preeminent if we look at shopping traffic patterns. A 2013 study from Shop.org, The Partnering Group, and comScore revealed that 55% of the time spent on retailers' websites was from mobile devices, compared to 45% from traditional computers. Clearly, while actual mobile sales are a significant driver for commerce, mobile shopping is even more important.

Finally, it is worth noting this stat, from Nielsen, which shows that 72% of smart-phone shoppers who make a purchase on their device do so at home, not "out and about". This is an important point when putting what we think of as mobile commerce into perspective. What the smartphone and other mobile Internet devices enable is more than just "out of home" commerce. What mobile technology makes possible is commerce wherever, and whenever, the buyer deems appropriate.

8.4 Beyond the Numbers

These statistics are important, but what do they mean? We've seen substantial growth rates for any number of technologies, channels, and platforms over the past few decades, but mobile is outpacing them all. Any kind of statistical analysis has to also recognize the rapid, dramatic rise in mobile-related behaviors. When we learn that 26% of mobile phone users have hooked up their phones to their cars to listen to media, this shows more than simply a typical adoption curve for new technology. It shows a clear, pent-up demand to engage in activities previously not possible, but certainly imagined. After all, it isn't necessarily a straightforward activity to connect your phone to a vehicle, and yet tens of millions of Americans have done just that to have more choice and control over the content they want to consume.

The rise in smartphone ownership, and the even more dramatic rise in mobile-related behaviors, is not just about technology. It's about enabling behaviors that are natural to humans, and there's no better way to think about them than to imagine a day in the life of a modern smartphone user.

First of all, how do you wake up in the morning? In 2013, Edison's Infinite Dial study asked that question of a representative sample of Americans aged 12 and older. The number one answer, at 30% of the population, was by setting an alarm on a mobile phone. Remember the movie *Groundhog Day*, in which Bill Murray's character kept waking up on the same calendar day to a radio morning show? Today, a plurality of Americans wake up to a *noise*—the noise of their smartphone alarm.

After shaking off the cobwebs of sleep, today's smartphone owner takes the phone off the nightstand and checks Facebook, the emails that came in the night before, the weather, and the news headlines. Statistically, smartphone owners check Facebook several more times throughout the day, because they *can*. But for now, a simple scan will do.

After breakfast, our mobile-savvy consumer gets dressed, packs his bag for the day, and turns once again to the mobile phone. For those who commute by car, a destination is loaded into a navigation app that features a real-time traffic subscription, informing of the fastest route to work. For those who take public transportation, apps are available that transmit the exact times that the bus or the subway will arrive, and the optimal path for the commuter to use to get in on time and with no wasted effort.

During the commute, either via car or public transportation, the smartphone user consumes media previously unavailable. Drivers listen to Pandora or a Spotify playlist. Bus riders listen to yesterday's NPR podcasts, or watch a news program on YouTube or Hulu. Some may even share a funny moment from those shows over Twitter, or snap a photo of their commute to post to Instagram, a behavior that was not possible just three years ago.

Before entering the office, our protagonist walks by a coffee shop. Just by passing through the doorway, a Near Field Communications (NFC) or Bluetooth Low-Energy

(BLE) chip notifies the coffee shop that a Cafe Americano is on order, and a connected mobile wallet takes care of the bill. Not a word is spoken as the exact order is placed, retrieved, and consumed.

That morning, our subject may work on a variety of tasks, online and off, but he takes a number of "digital vacations" at various times throughout the day. Many of these breaks take the form of checking Facebook or other social media platforms. At one point, our hero sees a friend post about a new music recording, or a book, or a movie. In that instant, a decision is made, and the book or the album, is seam-lessly purchased and downloaded to the phone.

Halfway through the day, our hero finishes downloading the new book or music and decides to go out to lunch. A location-based application finds a nearby restaurant, and another app secures a reservation without so much as a phone call.

During lunch, the smartphone is retrieved once again, to submit a review of the restaurant on Yelp, or simply to check in on Swarm to let friends know the best place for a hot dog downtown.

After lunch, on the walk back to the office, our hero hears about a new TV show and searches his phone for the details. After reading several positive reviews, he orders a few episodes, or even a whole season, to watch on the way home. A text is sent to a spouse: "Pizza and a movie?" "Good idea!" By the time the commute home rolls around, the movie is ready to go and dinner is ordered for pickup via a mobile app.

During the movie, a product placement shows up for a new sports car. Once again, the smartphone is employed, and several reviews of the car are found while sitting on the couch. A decision is made: A test-drive seems like a good idea. An appointment is made via email with a local dealer during the movie, and another app is consulted to line up possible financing.

Finally, it's time for bed. An alarm is set on the phone, tomorrow's weather consulted, and a conversation with a spouse points the way to a book that a son or daughter spoke about. The phone is brought out once again, and an order placed with an online retailer; the book will arrive in two days. The lights go out, and our hero goes to bed, to sleep the sleep of champions.

None of what you have just read was possible even five years ago. And yet, nothing here is science fiction or implausible—only new ways to do the things we've always done, like order pizza. That is the point of this chapter: The advances in mobile technology are not about enabling things that were previously unimaginable. These things were all imaginable. What mobile technology enables is the ability to do things where and when we want to do them, plain and simple, and that as much as anything has led to the dramatic rise in smartphone ownership and usage over the past three years. Smartphones make doing the things we already do even easier.

The current state of mobile does not enable some strange or foreign activity, but the ability to engage in the familiar, no matter where we are. In fact, the mobile commerce revolution is not about technology, but rather about what we can *do* with that technology, and how it enables and empowers us to engage in natural behaviors that we didn't even know we could engage in. If you travel to a new city and wonder where you should go for dinner, you've never had more information at your fingertips than you do right now, and mobile technology is a great equalizer in that sense. With near-perfect information available at our fingertips in terms of local business reviews, for instance, the best "mobile" strategy for a restaurant is to be a great restaurant, period. Thanks to mobile review and reservation apps, there's simply no other way to survive in a world with near-perfect, instant, real-time communication.

8.5 The Bottom Line

The advances in mobile technology over the past five years are unlike any other advances seen before. Some technological advances open our eyes to things we never dreamed we could do, or even had the language to describe. That's not what mobile technology is about. Instead, the statistics in this chapter underline what is truly powerful about mobile technology: It's not about enabling things we couldn't have imagined. Instead, today's mobile technology enables exactly what the average consumer *could* have imagined, albeit in settings and situations that could not have been predicted.

Mobile technology and mobile broadband in particular, enables us to consume media, purchase products, and, yes, settle bar bets in situations that were not possible a few years ago. In short, mobile enhances our everyday lives in ways that many of us now take for granted, which is truly remarkable for a technology that is only a few years old. We grew up watching Cliff on *Cheers* telling us about the Egyptians, how World War I started, or the history of capitalism. Today, in the great meritocracy of the mobile world, we *all* have access to that information. The bar bet is over. And the great mobile commerce revolution is just beginning.

8.6 New Words and Expressions

eponymous	*adj.*	使得名的,齐名的
demographic	*adj.*	人口统计学的;人口学的
psychographic	*adj.*	心理记录的
communicative	*adj.*	爱说话的,健谈的

retrospect	*n.* 回顾，追溯
intriguing	*adj.* 有趣的；迷人的
ITU(International Telecommunication Union)	国际电信联盟
shake off	摆脱；抖落
optimal path	最优路径
commuter	*n.* 通勤者，经常乘公共车辆往返者
protagonist	*n.* 主角，主演
NPR(National Public Radio)	美国国家公共电台
Americano	*n.* 美式咖啡
take for granted	认为是理所当然的

8.7　Questions

Single Choice Questions

1. What are the tens of millions of users of both Instagram and Snapchat using their phones to share in altogether new ways? (　　)
    A. Images
    B. Images and messages
    C. Images, messages, and above all, experiences
    D. None of the above

2. Which of the following statements is true? (　　)
    A. Much has been said about the "platform wars" between Google's Android, Apple's iOS and BlackBerry OS
    B. Much has been said about the "platform wars" between Google's Android, Microsoft's Windows Mobile and BlackBerry OS
    C. Much has been said about the "platform wars" between Apple's iOS, Microsoft's Windows Mobile and BlackBerry OS
    D. Much has been said about the "platform wars" between Google's Android, Apple's iOS, Microsoft's Windows Mobile, and BlackBerry OS

3. How many percent of growth rate for mobile retail commerce is significantly higher than the growth rate for mobile phone penetration? (　　)
    A. 61%　　　　B. 71%　　　　C. 81%　　　　D. 91%

4. We've seen substantial growth rates for any number of technologies, channels, platforms and mobile over the past few decades. Which is the most prominent representative? (　　)
    A. Technologies　　B. Channels　　C. Platforms　　D. Mobile

5. To what kind of noise a plurality of Americans wake up? (　　)
    A. Smartphone alarm　　　　B. Radio morning show

C. Television   D. People
6. Which of the following statements is true? (   )
   A. The current state of mobile enable both some strange or foreign activity and the ability to engage in the familiar, no matter where we are
   B. The current state of mobile does not enable some strange or foreign activity, but the ability to engage in the familiar, no matter where we are
   C. The current state of mobile enable some strange or foreign activity, but the ability to engage in the familiar, no matter where we are
   D. The current state of mobile neither enable some strange or foreign activity, nor the ability to engage in the familiar, no matter where we are
7. With mobile review and reservation apps, what kind of the way in which we have to survive in a world? (   )
   A. Near-perfect   B. Instan
   C. Real-time communication   D. Above of the above
8. Which of the following item do mobile technology and mobile broadband in particular enable us to do in situations that were not possible a few years ago? (   )
   A. Consume media and purchase products
   B. Consume media and settle bar bets
   C. Purchase products and settle bar bets
   D. Consume media, purchase products, and settle bar bets
9. Which of the following statements is true? (   )
   A. In short, mobile enhances our everyday lives in ways that many of us now take for granted, which is truly remarkable for a technology that is only a few months old
   B. In short, mobile enhances our everyday lives in ways that many of us now take for granted, which is truly remarkable for a technology that is only a few weeks old
   C. In short, mobile enhances our everyday lives in ways that many of us now take for granted, which is truly remarkable for a technology that is only a few years old
   D. In short, mobile enhances our everyday lives in ways that many of us now take for granted, which is truly remarkable for a technology that is only a few days old
10. Which of the following statements is true? (   )
    A. Today, in the great ergatocracy of the mobile world, we all have access to that information
    B. Today, in the great meritocracy of the mobile world, we all have access to that information

C. Today, in the great expertocracy of the mobile world, we all have access to that information

D. Today, in the great sciocracy of the mobile world, we all have access to that information

8.8　Problems

After reading this chapter and completing the exercises, you will be able to do the following:

1. What is the Mobile Commerce Revolution?
2. Explain the Current State of Mobile.
3. What is smartphone?
4. How to explain the Mobile around the World?
5. Discuss the mobile commerce.
6. Explain beyond the numbers and the bottom line.

Unit 9   Security—Information Security Principles of Success

9.1   Introduction

Many of the topics information technology students study in school carry directly from the classroom to the workplace. For example, new programming and systems analysis and design skills can often be applied on new systems-development projects as companies espouse cloud computing and mobile infrastructures that access internal systems.

Security is a little different. Although their technical skills are certainly important, the best security specialists combine their practical knowledge of computers and networks with general theories about security, technology, and human nature. These concepts, some borrowed from other fields, such as military defense, often take years of (sometimes painful) professional experience to learn. With a conceptual and principled view of information security, you can analyze a security need in the right frame of reference or context so you can balance the needs of permitting access against the risk of allowing such access. No two systems or situations are identical, and no cookbooks can specify how to solve certain security problems. Instead, you must rely on principle-based analysis and decision making.

This chapter introduces these key information security principles, concepts, and durable "truths."

9.2   Principle 1: There Is No Such Thing As Absolute Security

In 2003, the art collection of the Whitworth Gallery in Manchester, England, included three famous paintings by Van Gogh, Picasso, and Gauguin. Valued at more than \$7 million, the paintings were protected by closed-circuit television (CCTV), a series of alarm systems, and 24-hour rolling patrols. Yet in late April 2003, thieves broke into the museum, evaded the layered security system, and made off with the three masterpieces. Several days later, investigators discovered the paintings in a nearby public restroom along with a note from the thieves saying, "The intention was not to steal, only to highlight the woeful security."

The burglars' lesson translates to the information security arena and illustrates the first principle of information security (IS): Given enough time, tools, skills, and inclination, a malicious person can break through any security measure. This principle applies to the physical world as well and is best illustrated with an analogy of safes or vaults that businesses commonly use to protect their assets. Safes are rated according to their resistance to attacks using a scale that describes how long it could take a burglar to open them. They are divided into categories based on the level of protection they can deliver and the testing they undergo. Four common classes of safe ratings are B-Rate, C-Rate, UL TL-15, and UL TL-30:

- **B-Rate**: B-Rate is a catchall rating for any box with a lock on it. This rating describes the thickness of the steel used to make the lockbox. No actual testing is performed to gain this rating.
- **C-Rate**: This is defined as a variably thick steel box with a 1-inch-thick door and a lock. No tests are conducted to provide this rating, either.
- **UL TL-15**: Safes with an Underwriters Laboratory (UL) TL-15 rating have passed standardized tests as defined in UL Standard 687 using tools and an expert group of safe-testing engineers. The UL TL-15 label requires that the safe be constructed of 1-inch solid steel or equivalent. The label means that the safe has been tested for a networking time of 15 minutes using "common hand tools, drills, punches hammers, and pressure applying devices." *Networking time* means that when the tool comes off the safe, the clock stops. Engineers exercise more than 50 different types of attacks that have proven effective for safecracking.
- **UL TL-30**: UL TL-30 testing is essentially the same as the TL-15 testing, except for the networking time. Testers get 30 minutes and a few more tools to help them gain access. Testing engineers usually have a safe's manufacturing blueprints and can disassemble the safe before the test begins to see how it works.

FYI: Confidentiality by another Name

Confidentiality is sometimes referred to as the principle of least privilege, meaning that users should be given only enough privilege to perform their duties, and no more. Some other synonyms for confidentiality you might encounter include *privacy*, *secrecy*, *and discretion*.

"Security Architecture and Design," security testing of hardware and software systems employs many of the same concepts of safe testing, using computers and custom-developed testing software instead of tools and torches. The outcomes of this testing are the same, though: As with software, no safe is burglar proof; security measures simply buy time. Of course, buying time is a powerful tool. Resisting attacks long enough provides the opportunity to catch the attacker in the act and to quickly recover from the incident. This leads to the second principle.

FYI: Confidentiality Models

Confidentiality models are primarily intended to ensure that no unauthorized access to information is permitted and that accidental disclosure of sensitive information is not possible. Common confidentiality controls are user IDs and passwords.

9.3 Principle 2: The Three Security Goals Are Confidentiality, Integrity, and Availability

All information security measures try to address at least one of three goals:
- Protect the confidentiality of data
- Preserve the integrity of data
- Promote the availability of data for authorized use

These goals form the confidentiality, integrity, availability (CIA) triad, the basis of all security programs (see Figure 9.1). Information security professionals who create policies and procedures (often referred to as governance models) must consider each goal when creating a plan to protect a computer system.

Figure 9.1   The CIA triad.

FYI: CIA Triad

The principle of information security protection of confidentiality, integrity, and availability cannot be overemphasized: This is central to all studies and practices in IS. You'll often see the term *CIA triad* to illustrate the overall goals for IS throughout the research, guidance, and practices you encounter.

9.3.1 Integrity Models

Integrity models keep data pure and trustworthy by protecting system data from intentional or accidental changes. Integrity models have three goals:
- Prevent unauthorized users from making modifications to data or programs
- Prevent authorized users from making improper or unauthorized modifications
- Maintain internal and external consistency of data and programs

An example of integrity checks is balancing a batch of transactions to make sure that all the information is present and accurately accounted for.

9.3.2 Availability Models

Availability models keep data and resources available for authorized use, especially during emergencies or disasters. Information security professionals usually address three common challenges to availability:

- Denial of service (DoS) due to intentional attacks or because of undiscovered flaws in implementation (for example, a program written by a programmer who is unaware of a flaw that could crash the program if a certain unexpected input is encountered).
- Loss of information system capabilities because of natural disasters (fires, floods, storms, or earthquakes) or human actions (bombs or strikes).
- Equipment failures during normal use.

Some activities that preserve confidentiality, integrity, and/or availability are granting access only to authorized personnel, applying encryption to information that will be sent over the Internet or stored on digital media, periodically testing computer system security to uncover new vulnerabilities, building software defensively, and developing a disaster recovery plan to ensure that the business can continue to exist in the event of a disaster or loss of access by personnel.

9.4 Principle 3: Defense in Depth as Strategy

A bank would never leave its assets inside an unguarded safe alone. Typically, access to the safe requires passing through layers of protection that might include human guards and locked doors with special access controls. Furthermore, the room where the safe resides could be monitored by closed-circuit television, motion sensors, and alarm systems that can quickly detect unusual activity. The sound of an alarm might trigger the doors to automatically lock, the police to be notified, or the room to fill with tear gas.

Layered security, as in the previous example, is known as defense in depth. This security is implemented in overlapping layers that provide the three elements needed to secure assets: prevention, detection, and response. Defense in depth also seeks to offset the weaknesses of one security layer by the strengths of two or more layers.

In the information security world, defense in depth requires layering security devices in a series that protects, detects, and responds to attacks on systems. For example, a typical Internet-attached network designed with security in mind includes routers, firewalls, and intrusion detection systems (IDS) to protect the network from would-be intruders; employs traffic analyzers and real-time human monitors who watch for anomalies as the network is being used to detect any breach in the layers of protection; and relies on automated mechanisms to turn off access or remove the system from the network in response to the detection of an intruder.

Finally, the security of each of these mechanisms must be thoroughly tested before deployment to ensure that the integrated system is suitable for normal operations. After all, a chain is only as good as its weakest link.

In Practice: Phishing for Dollars

Phishing is another good example of how easily intelligent people can be duped into breaching security. Phishing is a dangerous Internet scam, and is becoming increasingly dangerous as targets are selected using data available from social media and enable a malicious person to build a profile of the target to better convince him the scam is real. A phishing scam typically operates as follows:

- The victim receives an official-looking email message purporting to come from a trusted source, such as an online banking site, PayPal, eBay, or other service where money is exchanged, moved, or managed.
- The email tells the user that his or her account needs updating immediately or will be suspended within a certain number of days.
- The email contains a URL (link) and instructs the user to click on the link to access the account and update the information. The link text appears as though it will take the user to the expected site. However, the link is actually a link to the attacker's site, which is made to look exactly like the site the user expects to see.
- At the spoofed site, the user enters his or her credentials (ID and password) and clicks Submit.
- The site returns an innocuous message, such as "We're sorry—we're unable to process your transaction at this time," and the user is none the wiser.
- At this point, the victim's credentials are stored on the attacker's site or sent via email to the perpetrator, where they can be used to log in to the *real* banking or exchange site and empty the account before the user knows what happened.

Phishing and resultant ID theft and monetary losses are on the increase and will begin to slow only after the cycle is broken through awareness and education. Protect yourself by taking the following steps:

- Look for telltale signs of fraud: Instead of addressing you by name, a phishing email addresses you as "User" or by your email address; a legitimate message from legitimate companies uses your name as they know it.
- Do not click on links embedded in unsolicited finance-related email messages. A link might look legitimate, but when you click on it, you could be redirected

to the site of a phisher. If you believe that your account is in jeopardy, type in the known URL of the site in a new browser window and look for messages from the provider after you're logged in.
- Check with your provider for messages related to phishing scams that the company is aware of. Your bank or other financial services provider wants to make sure you don't fall victim and will often take significant measures to educate users on how to prevent problems.

9.5 Principle 4: When Left on Their Own, People Tend to Make the Worst Security Decisions

The primary reason identity theft, viruses, worms, and stolen passwords are so common is that people are easily duped into giving up the secrets technologies use to secure systems. Organizers of Infosecurity Europe, Britain's biggest information technology security exhibition, sent researchers to London's Waterloo Station to ask commuters to hand over their office computer passwords in exchange for a free pen. Three-quarters of respondents revealed the information immediately, and an additional 15 percent did so after some gentle probing. Study after study like this one shows how little it takes to convince someone to give up their credentials in exchange for trivial or worthless goods.

9.6 Principle 5: Computer Security Depends on Two Types of Requirements: Functional and Assurance

Functional requirements describe what a system *should* do. Assurance requirements describe how functional requirements should be implemented and tested. Both sets of requirements are needed to answer the following questions:
- Does the system do the right things (behave as promised)?
- Does the system do the right things in the right way?

These are the same questions that others in noncomputer industries face with verification and validation. Verification is the process of confirming that one or more predetermined requirements or specifications are met. Validation then determines the correctness or quality of the mechanisms used to meet the needs. In other words, you can develop software that addresses a need, but it might contain flaws that could compromise data when placed in the hands of a malicious user.

Consider car safety testing as an example. Verification testing for seat belt functions might include conducting stress tests on the fabric, testing the locking mechanisms, and making certain the belt will fit the intended application, thus completing the functional tests. Validation, or assurance testing, might then include crashing the car with crash-test dummies inside to "prove" that the seat belt is indeed safe when used under normal conditions and that it can survive under harsh conditions.

With software, you need both verification and validation answers to gain confidence in products before launching them into a wild, hostile environment such as the Internet. Most of today's commercial off-the-shelf (COTS) software and systems stop at the first step, verification, without bothering to test for obvious security vulnerabilities in the final product. Developers of software generally lack the wherewithal and motivation needed to try to break their own software. More often, developers test that the software meets the specifications in each function that is present but usually do not try to find ways to circumvent the software and make it fail.

9.7 Principle 6: Security Through Obscurity Is Not an Answer

Many people in the information security industry believe that if malicious attackers don't know how software is secured, security is better. Although this might seem logical, it's actually untrue. Security through obscurity means that hiding the details of the security mechanisms is sufficient to secure the system alone. An example of security through obscurity might involve closely guarding the written specifications for security functions and preventing all but the most trusted people from seeing it. Obscuring security leads to a false sense of security, which is often more dangerous than not addressing security at all.

If the security of a system is maintained by keeping the implementation of the system a secret, the entire system collapses when the first person discovers how the security mechanism works—and someone is always determined to discover these secrets. The better bet is to make sure no one mechanism is responsible for the security of the entire system. Again, this is defense in depth in everything related to protecting data and resources.

"Cryptography," you'll see how this principle applies and why it makes no sense to keep an algorithm for cryptography secret when the security of the system should rely on the cryptographic keys used to protect data or authenticate a user. You can also see this in action with the open-source movement: Anyone can gain access to program (source) code, analyze it for security problems, and then share with the community improvements that eliminate vulnerabilities and/or improve the overall security through simplification (see Principle 9).

9.8　Principle 7: Security = Risk Management

It's critical to understand that spending more on securing an asset than the intrinsic value of the asset is a waste of resources. For example, buying a $500 safe to protect $200 worth of jewelry makes no practical sense. The same is true when protecting electronic assets. All security work is a careful balance between the level of risk and the expected reward of expending a given amount of resources. Security is concerned not with eliminating all threats within a system or facility, but with eliminating known threats and minimizing losses if an attacker succeeds in exploiting a vulnerability. Risk analysis and risk management are central themes to securing information systems. When risks are well understood, three outcomes are possible:

- The risks are mitigated (countered).
- Insurance is acquired against the losses that would occur if a system were compromised.
- The risks are accepted and the consequences are managed.

Risk assessment and risk analysis are concerned with placing an economic value on assets to best determine appropriate countermeasures that protect them from losses. The simplest form of determining the degree of a risk involves looking at two factors:

- What is the consequence of a loss?
- What is the likelihood that this loss will occur?

Figure 9.2 illustrates a matrix you can use to determine the degree of a risk based on these factors.

Likelihood	Consequences				
	1.Insignificant	2.Minor	3.Moderate	4.Major	6.Catastrophic
A(almost certain)	High	High	Extreme	Extreme	Extreme
B(likely)	Moderate	High	High	Extreme	Extreme
C(moderate)	Low	Moderate	High	Extreme	Extreme
D(unlikely)	Low	Low	Moderate	High	Extreme
E(rare)	Low	Low	Moderate	High	High

Figure 9.2　Consequences/likelihood matrix for risk analysis.

After determining a risk rating, one of the following actions could be required:

- **Extreme risk**: Immediate action is required.
- **High risk**: Senior management's attention is needed.
- **Moderate risk**: Management responsibility must be specified.
- **Low risk**: Management is handled by routine procedures.

In the real world, risk management is more complicated than simply making a human judgment call based on intuition or previous experience with a similar situation. Recall that every system has unique security issues and considerations, so it's imperative to understand the specific nature of data the system will maintain, what hardware and software will be used to deploy the system, and the security skills of the development teams. Determining the likelihood of a risk coming to life requires understanding a few more terms and concepts:

- Vulnerability.
- Exploit.
- Attacker.

Vulnerability refers to a known problem within a system or program. A common example in InfoSec is called the buffer overflow or buffer overrun vulnerability. Programmers tend to be trusting and not worry about who will attack their programs, but instead worry about who will use their programs legitimately. One feature of most programs is the capability for a user to "input" information or requests. The program instructions (source code) then contain an "area" in memory (buffer) for these inputs and act upon them when told to do so. Sometimes the programmer doesn't check to see if the input is proper or innocuous. A malicious user, however, might take advantage of this weakness and overload the input area with more information than it can handle, crashing or disabling the program. This is called buffer overflow, and it can permit a malicious user to gain control over the system. This common vulnerability with software must be addressed when developing systems.

An exploit is a program or "cookbook" on how to take advantage of a specific vulnerability. It might be a program that a hacker can download over the Internet and then use to search for systems that contain the vulnerability it's designed to exploit. It might also be a series of documented steps on how to exploit the vulnerability after an attacker finds a system that contains it.

An attacker, then, is the link between a vulnerability and an exploit. The attacker has two characteristics: skill and will. Attackers either are skilled in the art of attacking systems or have access to tools that do the work for them. They have the will to perform attacks on systems they do not own and usually care little about the consequences of their actions.

In applying these concepts to risk analysis, the IS practitioner must anticipate who might want to attack the system, how capable the attacker might be, how available the exploits to a vulnerability are, and which systems have the vulnerability present.

Risk analysis and risk management are specialized areas of study and practice, and the IS professionals who concentrate in these areas must be skilled and current in their techniques.

9.9 Principle 8: The Three Types of Security Controls Are Preventative, Detective, and Responsive

Controls (such as documented processes) and countermeasures (such as firewalls) must be implemented as one or more of these previous types, or the controls are not there for the purposes of security. Shown in another triad, the principle of defense in depth dictates that a security mechanism serve a purpose by preventing a compromise, detecting that a compromise or compromise attempt is underway, or responding to a compromise while it's happening or after it has been discovered.

Referring to the example of the bank vault in Principle 3, access to a bank's safe or vault requires passing through layers of protection that might include human guards and locked doors with special access controls (prevention). In the room where the safe resides, closed-circuit televisions, motion sensors, and alarm systems quickly detect any unusual activity (detection). The sound of an alarm could trigger the doors to automatically lock, the police to be notified, or the room to fill with tear gas (response).

These controls are the basic toolkit for the security practitioner who mixes and matches them to carry out the objectives of confidentiality, integrity, and/or availability by using people, processes, or technology (see Principle 11) to bring them to life.

In Practice: How People, Process, and Technology Work in Harmony

To illustrate how people, process, and technology work together to secure systems, let's take a look a how the security department grants access to users for performing their duties. The process, called user access request, is initiated when a new user is brought into the company or switches department or role within the company. The user access request form is initially completed by the user and approved by the manager.

When the user access request is approved, it's routed to information security access coordinators to process using the documented procedures for granting access. After access is granted and the process for sharing the user's ID and password is followed, the system's technical access control system takes over. It protects the system from unauthorized access by requiring a user ID and password, and it prevents password guessing from an unauthorized person by limiting the number of attempts to three before locking the account from further access attempts.

In Practice: To Disclose or Not to Disclose—That Is the Question!

Having specific knowledge of a security vulnerability gives administrators the knowledge to properly defend their systems from related exploits. The ethical question is, how should that valuable information be disseminated to the good guys while keeping it away from the bad guys? The simple truth is, you can't really do this. Hackers tend to communicate among themselves far better than professional security practitioners ever could. Hackers know about most vulnerabilities long before the general public gets wind of them. By the time the general public is made aware, the hacker community has already developed a workable exploit and disseminated it far and wide to take advantage of the flaw before it can be patched or closed down.

Because of this, open disclosure benefits the general public far more than is acknowledged by the critics who claim that it gives the bad guys the same information.

Here's the bottom line: If you uncover an obvious problem, raise your hand and let someone who can do something about it know. If you see something, say something. You'll sleep better at night!

9.10　Principle 9: Complexity Is the Enemy of Security

The more complex a system gets, the harder it is to secure. With too many "moving parts" or interfaces between programs and other systems, the system or interfaces become difficult to secure while still permitting them to operate as intended.

9.11　Principle 10: Fear, Uncertainty, and Doubt Do Not Work in Selling Security

At one time, "scaring" management into spending resources on security to avoid the unthinkable was effective. The tactic of fear, uncertainty, and doubt (FUD) no longer works: Information security and IT management is too mature. Now IS managers must justify all investments in security using techniques of the trade. Although this makes the job of information security practitioners more difficult, it also makes them more valuable because of management's need to understand what is being protected and why. When spending resources can be justified with good, solid business rationale, security requests are rarely denied.

9.12　Principle 11: People, Process, and Technology Are All Needed to Adequately Secure a System or Facility

As described in Principle 3, "Defense in Depth as Strategy," the information security practitioner needs a series of countermeasures and controls to implement an effective security system. One such control might be dual control, a practice borrowed from the military. The U.S. Department of Defense uses a dual control protocol to secure the nation's nuclear arsenal. This means that at least two on-site people must agree to launch a nuclear weapon. If one person were in control, he or she could make an error in judgment or act maliciously for whatever reason. But with dual control, one person acts as a countermeasure to the other: Chances are less likely that both people will make an error in judgment or act maliciously. Likewise, no one person in an organization should have the ability to control or close down a security activity. This is commonly referred to as separation of duties.

Process controls are implemented to ensure that different people can perform the same operations exactly in the same way each time. Processes are documented as procedures on how to carry out an activity related to security. The process of configuring a server operating system for secure operations is documented as one or more procedures that security administrators use and can be verified as done correctly.

Just as the information security professional might establish process controls to make sure that a single person cannot gain complete control over a system, you should never place all your faith in technology. Technology can fail, and without people to notice and fix technical problems, computer systems would stall permanently. An example of this type of waste is installing an expensive firewall system (a network perimeter security device that blocks traffic) and then turning around and opening all the ports that are intended to block certain traffic from entering the network.

People, process, and technology controls are essential elements of several areas of practice in information technology (IT) security, including operations security, applications development security, physical security, and cryptography. These three pillars of security are often depicted as a three-legged stool (see Figure 9.3).

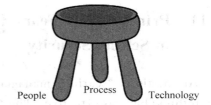

Figure 9.3 The people, process, and technology triad.

9.13 Principle 12: Open Disclosure of Vulnerabilities Is Good for Security!

A raging and often heated debate within the security community and software developing centers concerns whether to let users know about a problem before a fix or patch can be developed and distributed. Principle 6 tells us that security through obscurity is not an answer: Keeping a given vulnerability secret from users and from the software developer can only lead to a false sense of security. Users have a right to know about defects in the products they purchase, just as they have a right to know about automobile recalls because of defects. The need to know trumps the need to keep secrets, to give users the right to protect themselves.

9.14 Summary

To be most effective, computer security specialists not only must know the technical side of their jobs, but also must understand the principles behind information security. No two situations that security professionals review are identical, and there are no recipes or cookbooks on universal security measures. Because each situation calls for a distinct judgment to address the specific risks inherent in information systems, principles-based decision making is imperative. An old saying goes, "If you only have a hammer, every problem looks like a nail." This approach simply does not serve today's businesses, which are always striving to balance risk and reward of access to electronic records. The goal is to help you create a toolkit and develop the skills to use these tools like a master craftsman. Learn these principles and take them to heart, and you'll start out much further along than your peers who won't take the time to bother learning them!

As you explore the rest of the Common Body of Knowledge (CBK) domains, try to relate the practices you find to one or more of these. This helps prevent breaches in confidentiality, integrity, and availability, and implements the principle of defense in depth. As you will find, these principles are mixed and matched to describe why certain security functions and operations exist in the real world of IT.

9.15 New Words and Expressions

evade	vt. 逃避;规避;逃脱
masterpiece	n. 杰作;绝无仅有的人
woeful	adj. 悲惨的;遗憾的
arena	n. 舞台;竞技场
malicious	adj. 恶意的;蓄意的
analogy	n. 类比;类推
confidentiality	n. 机密性
principle of least privilege	最少特权原则
synonym	n. 同义词;同义字
IDS(intrusion detection systems)	入侵检测系统
Anomaly	n. 异常现象,反常现象
phishing	n. 网络钓鱼;网络欺诈

scam	*n.*	骗局,诡计
victim	*n.*	受害人;牺牲品
paypal	*n.*	贝宝(全球最大的在线支付平台)
eBay	*n.*	易趣(知名网上购物网站)
spoof	*vt.*	哄骗;戏弄
perpetrator	*n.*	犯罪者;行凶者
legitimate	*adj.*	合法的;正当的
cryptography	*n.*	密码学;密码使用法
vulnerability	*n.*	易损性;弱点
countermeasure	*n.*	对策;防治对策
imperative	*adj.*	必要的;紧急的
CBK(Common Body of Knowledge)		公共知识体系

9.16 Questions

Single Choice Questions

1. Which one is not one of the three goals that all information security measures try to address? (   )

    A. Protect the confidentiality of data

    B. Protect a computer system

    C. Preserve the integrity of data

    D. Promote the availability of data for authorized use

2. Which of the following statement is not safe rating? (   )

    A. UL TL-25   B. B-Rate   C. C-Rate   D. UL TL-30

3. Which of the following statement is not integrity models' goal? (   )

    A. Prevent unauthorized users from making modifications to data or programs

    B. Prevent authorized users from making improper or unauthorized modifications

    C. Keep data pure

    D. Maintain internal and external consistency of data and programs

4. Which of the following sentence describes the two types of requirements? (   )

    A. Functional and logical   B. Logical and physical

    C. Functional and assurance   D. Functional and physical

5. After determining a risk rating, which one of the following actions could not be required? (   )

    A. Extreme risk   B. High risk   C. General risk   D. Low risk

6. Which of the following represents the three types of security controls? (   )

    A. Preventative   B. Detective   C. Responsive   D. All of them

7. The CIA triad is often represented by which of the following? (   )

   A. Confidentiality   B. Integrity   C. Availability   D. All of them

8. Which of the following terms best describes the assurance that data has not been changed unintentionally due to an accident or malice? (   )

   A. The more simple a system gets, the harder it is to secure

   B. The more complex a system gets, the easier it is to secure

   C. The more complex a system gets, the harder it is to secure

   D. None of above

9. Which are essential elements of several areas of practice in Information Technology (IT) security? (   )

   A. People, functions and technology

   B. People, process, and technology controls

   C. Technology, roles and separation of duties

   D. Separation of duties, processes and people

10. Which one is not required on determining the likelihood of a risk coming to life? (   )

   A. Vulnerability   B. Exploit   C. Attacker   D. Speed

9.17 Problems

After reading this chapter and completing the exercises, you will be able to do the following:

1. Distinguish among the three main security goals.

2. Learn how to design and apply the principle of defense in depth.

3. Comprehend human vulnerabilities in security systems to better design solutions to counter them.

4. Explain the difference between functional requirements and assurance requirements.

5. Determine which side of the open disclosure debate you would take.

6. Evaluating Real-World Defense in Depth.

Unit 10 Web Services—Designing Software in a Distributed World

This chapter is an overview of what is involved in designing services that use distributed computing techniques. These are the techniques all large web sites use to achieve their size, scale, speed, and reliability.

There are two ways of constructing a software design: One way is to make it so simple that there are obviously no deficiencies and the other way is to make it so complicated that there are no obvious deficiencies.

—C.A.R. Hoare, The 1980 ACM Turing Award Lecture

How does Google Search work? How does your Facebook Timeline stay updated around the clock? How does Amazon scan an ever-growing catalog of items to tell you that people who bought this item also bought socks?

Is it magic? No, it's distributed computing.

This chapter is an overview of what is involved in designing services that use distributed computing techniques. These are the techniques all large web sites use to achieve their size, scale, speed, and reliability.

Distributed computing is the art of building large systems that divide the work over many machines. Contrast this with traditional computing systems where a single computer runs software that provides a service, or client-server computing where many machines remotely access a centralized service. In distributed computing there are typically hundreds or thousands of machines working together to provide a large service.

Distributed computing is different from traditional computing in many ways. Most of these differences are due to the sheer size of the system itself. Hundreds or thousands of computers may be involved. Millions of users may be served. Billions and sometimes trillions of queries may be processed.

Terms to Know

Server: Software that provides a function or application program interface (API). (Not a piece of hardware.)

Service: A user-visible system or product composed of many servers.

Machine: A virtual or physical machine.

QPS: Queries per second. Usually how many web hits or API calls received per second.

Traffic: A generic term for queries, API calls, or other requests sent to a server.

Performant: A system whose performance conforms to (meets or exceeds) the design requirements. A neologism from merging "performance" and "conformant."

Application Programming Interface (API): A protocol that governs how one server talks to another.

Speed is important. It is a competitive advantage for a service to be fast and responsive. Users consider a web site sluggish if replies do not come back in 200 ms or less. Network latency eats up most of that time, leaving little time for the service to compose the page itself.

In distributed systems, failure is normal. Hardware failures that are rare, when multiplied by thousands of machines, become common. Therefore failures are assumed, designs work around them, and software anticipates them. Failure is an expected part of the landscape.

Due to the sheer size of distributed systems, operations must be automated. It is inconceivable to manually do tasks that involve hundreds or thousands of machines. Automation becomes critical for preparation and deployment of software, regular operations, and handling failures.

10.1 Visibility at Scale

To manage a large distributed system, one must have visibility into the system. The ability to examine internal state—called **introspection**—is required to operate, debug, tune, and repair large systems.

In a traditional system, one could imagine an engineer who knows enough about the system to keep an eye on all the critical components or "just knows" what is wrong based on experience. In a large system, that level of visibility must be actively created by designing systems that draw out the information and make it visible. No person or team can manually keep tabs on all the parts.

Distributed systems, therefore, require components to generate copious logs that detail what happened in the system. These logs are then aggregated to a central location for collection, storage, and analysis. Systems may log information that is very high level, such as whenever a user makes a purchase, for each web query, or for every API call. Systems may log low-level information as well, such as the parameters of every function call in a critical piece of code.

Systems should export metrics. They should count interesting events, such as how many times a particular API was called, and make these counters accessible.

In many cases, special URLs can be used to view this internal state. For example, the Apache HTTP Web Server has a "server-status" page (http://www.example.com/server-status/).

In addition, components of distributed systems often appraise their own health and make this information visible. For example, a component may have a URL that outputs whether the system is ready (OK) to receive new requests. Receiving as output anything other than the byte "O" followed by the byte "K" (including no response at all) indicates that the system does not want to receive new requests. This information is used by load balancers to determine if the server is healthy and ready to receive traffic. The server sends negative replies when the server is starting up and is still initializing, and when it is shutting down and is no longer accepting new requests but is processing any requests that are still in flight.

10.2 The Importance of Simplicity

It is important that a design remain as simple as possible while still being able to meet the needs of the service. Systems grow and become more complex over time. Starting with a system that is already complex means starting at a disadvantage.

Providing competent operations requires holding a mental model of the system in one's head. As we work we imagine the system operating and use this mental model to track how it works and to debug it when it doesn't. The more complex the system, the more difficult it is to have an accurate mental model. An overly complex system results in a situation where no single person understands it all at any one time.

In *The Elements of Programming Style*, Kernighan and Plauger (1978) wrote:

Debugging is twice as hard as writing the code in the first place. Therefore, if you write the code as cleverly as possible, you are, by definition, not smart enough to debug it.

The same is true for distributed systems. Every minute spent simplifying a design pays off time and time again when the system is in operation.

10.3 Composition

Distributed systems are composed of many smaller systems. In this section, we explore three fundamental composition patterns in detail:

- Load balancer with multiple backend replicas
- Server with multiple backends
- Server tree

10.3.1 Load Balancer with Multiple Backend Replicas

The first composition pattern is the load balancer with multiple backend replicas.

As depicted in Figure 10.1, requests are sent to the load balancer server. For each request, it selects one **backend** and forwards the request there. The response comes back to the load balancer server, which in turn relays it to the original requester.

The backends are called **replicas** because they are all clones or replications of each other. A request sent to any replica should produce the same response.

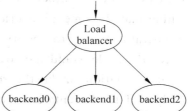

Figure 10.1 A load balancer with many replicas.

The load balancer must always know which backends are alive and ready to accept requests. Load balancers send **health check** queries dozens of times each second and stop sending traffic to that backend if the health check fails. A health check is a simple query that should execute quickly and return whether the system should receive traffic.

Picking which backend to send a query to can be simple or complex. A simple method would be to alternate among the backends in a loop—a practice called **round-robin**. Some backends may be more powerful than others, however, and may be selected more often using a proportional round-robin scheme. More complex solutions include the **least loaded** scheme. In this approach, a load balancer tracks how loaded each backend is and always selects the least loaded one.

Selecting the least loaded backend sounds reasonable but a naive implementation can be a disaster. A backend may not show signs of being overloaded until long after it has actually become overloaded. This problem arises because it can be difficult to accurately measure how loaded a system is. If the load is a measurement of the number of connections recently sent to the server, this definition is blind to the fact that some connections may be long lasting while others may be quick. If the measurement is based on CPU utilization, this definition is blind to input/output (I/O) overload. Often a trailing average of the last 5 minutes of load is used. Trailing averages have a problem in that, as an average, they reflect the past, not the present. As a consequence, a sharp, sudden increase in load will not be reflected in the average for a while.

Imagine a load balancer with 10 backends. Each one is running at 80 percent load. A new backend is added. Because it is new, it has no load and, therefore, is the least loaded backend. A naive least loaded algorithm would send all traffic to this new backend; no traffic would be sent to the other 10 backends. All too quickly, the new backend would become absolutely swamped. There is no way a single backend could process the traffic previously handled by 10 backends. The use of trailing averages would mean the older backends would continue reporting artificially high loads for a few minutes while the new backend would be reporting an artificially low load.

With this scheme, the load balancer will believe that the new machine is less loaded than all the other machines for quite some time. In such a situation the machine may become so overloaded that it would crash and reboot, or a system administrator trying to rectify the situation might reboot it. When it returns to service, the cycle would start over again.

Such situations make the round-robin approach look pretty good. A less naive least loaded implementation would have some kind of control in place that would never send more than a certain number of requests to the same machine in a row. This is called a **slow start** algorithm.

Trouble with a Naive Least Loaded Algorithm

Without slow start, load balancers have been known to cause many problems. One famous example is what happened to the CNN.com web site on the day of the September 11, 2001, terrorist attacks. So many people tried to access CNN.com that the backends became overloaded. One crashed, and then crashed again after it came back up, because the naive least loaded algorithm sent all traffic to it. When it was down, the other backends became overloaded and crashed. One at a time, each backend would get overloaded, crash, and become overloaded from again receiving all the traffic and crash again.

As a result the service was essentially unavailable as the system administrators rushed to figure out what was going on. In their defense, the web was new enough that no one had experience with handling sudden traffic surges like the one encountered on September 11.

The solution CNN used was to halt all the backends and boot them at the same time so they would all show zero load and receive equal amounts of traffic.

The CNN team later discovered that a few days prior, a software upgrade for their load balancer had arrived but had not yet been installed. The upgrade added a slow start mechanism.

10.3.2 Server with Multiple Backends

The next composition pattern is a server with multiple backends. The server receives a request, sends queries to many backend servers, and composes the final reply by combining those answers. This approach is typically used when the original query can easily be deconstructed into a number of independent queries that can be combined to form the final answer.

Figure 10.2(a) illustrates how a simple search engine processes a query with the help of multiple backends. The frontend receives the request. It relays the query to many backend servers. The spell checker replies with information so the search engine may suggest alternate spellings. The web and image search backends reply with a list of web sites and images related to the query. The advertisement server replies with advertisements relevant to the query. Once the replies are received, the frontend uses this information to construct the HTML that makes up the search results page for the user, which is then sent as the reply.

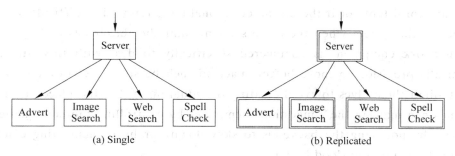

Figure 10.2   This service is composed of a server and many backends.

Figure 10.2(b) illustrates the same architecture with replicated, load-balanced, backends. The same principle applies but the system is able to scale and survive failures better.

This kind of composition has many advantages. The backends do their work in parallel. The reply does not have to wait for one backend process to complete before the next begins. The system is loosely coupled. One backend can fail and the page can still be constructed by filling in some default information or by leaving that area blank.

This pattern also permits some rather sophisticated latency management. Suppose this system is expected to return a result in 200 ms or less. If one of the backends is slow for some reason, the frontend doesn't have to wait for it. If it takes 10 ms to compose and send the resulting HTML, at 190 ms the frontend can give up on the slow backends and generate the page with the information it has. The ability to manage a latency time budget like that can be very powerful. For example, if the advertisement system is slow, search results can be displayed without any ads.

To be clear, the terms "frontend" and "backend" are a matter of perspective. The frontend sends requests to backends, which reply with a result. A server can be both a frontend and a backend. In the previous example, the server is the backend to the web browser but a frontend to the spell check server.

There are many variations on this pattern. Each backend can be replicated for increased capacity or resiliency. Caching may be done at various levels.

The term **fan out** refers to the fact that one query results in many new queries, one to each backend. The queries "fan out" to the individual backends and the replies **fan** in as they are set up to the frontend and combined into the final result.

Any fan in situation is at risk of having congestion problems. Often small queries may result in large responses. Therefore a small amount of bandwidth is used to fan out but there may not be enough bandwidth to sustain the fan in. This may result in congested network links and overloaded servers. It is easy to engineer the system to have the right amount of network and server capacity if the sizes of the queries and replies are consistent, or if there is an occasional large reply. The difficult situation is engineering the system when there are sudden, unpredictable bursts of large replies. Some network equipment is engineered specifically to deal with this situation by dynamically provisioning more buffer space to such bursts. Likewise, the backends can rate-limit themselves to avoid creating the situation in the first place. Lastly, the frontends can manage the congestion themselves by controlling the new queries they send out, by notifying the backends to slow down, or by implementing emergency measures to handle the flood better.

10.3.3 Server Tree

The other fundamental composition pattern is the **server tree**. As Figure 10.3 illustrates, in this scheme a number of servers work cooperatively with one as the root of the tree, parent servers below it, and leaf servers at the bottom of the tree. (In computer science, trees are drawn upside-down.) Typically this pattern is used to access a large dataset or **corpus**. The corpus is larger than any one machine can hold; thus each leaf stores one fraction or **shard** of the whole.

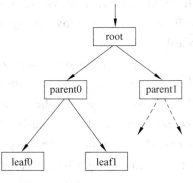

Figure 10.3 A server tree.

To query the entire dataset, the root receives the original query and forwards it to the parents. The parents forward the query to the leaf servers, which search their parts of the corpus. Each leaf sends its findings to the parents, which sort and filter the results before forwarding them up to the root. The root then takes the response from all the parents, combines the results, and replies with the full answer.

Imagine you wanted to find out how many times George Washington was mentioned in an encyclopedia. You could read each volume in sequence and arrive at the answer. Alternatively, you could give each volume to a different person and have the various individuals search their volumes in parallel. The latter approach would complete the task much faster.

The primary benefit of this pattern is that it permits parallel searching of a large corpus. Not only are the leaves searching their share of the corpus in parallel, but the sorting and ranking performed by the parents are also done in parallel.

For example, imagine a corpus of the text extracted from every book in the U.S. Library of Congress. This cannot fit in one computer, so instead the information is spread over hundreds or thousands of leaf machines. In addition to the leaf machines are the parents and the root. A search query would go to a root server, which in turn relays the query to all parents. Each parent repeats the query to all leaf nodes below it. Once the leaves have replied, the parent ranks and sorts the results by relevancy.

For example, a leaf may reply that all the words of the query exist in the same paragraph in one book, but for another book only some of the words exist (less relevant), or they exist but not in the same paragraph or page (even less relevant). If the query is for the best 50 answers, the parent can send the top 50 results to the root and drop the rest. The root then receives results from each parent and selects the best 50 of those to construct the reply.

This scheme also permits developers to work within a latency budget. If fast answers are more important than perfect answers, parents and roots do not have to wait for slow replies if the latency deadline is near.

Many variations of this pattern are possible. Redundant servers may exist with a load-balancing scheme to divide the work among them and route around failed servers. Expanding the number of leaf servers can give each leaf a smaller portion of the corpus to search, or each shard of corpus can be placed on multiple leaf servers to improve availability. Expanding the number of parents at each level increases the capacity to sort and rank results. There may be additional levels of parent servers, making the tree taller. The additional levels permit a wider fanout, which is important for an extremely large corpus. The parents may provide a caching function to relieve pressure on the leaf servers; in this case more levels of parents may improve cache effectiveness. These techniques can also help mitigate congestion problems related to fan-in, as discussed in the previous section.

10.4 Distributed State

Large systems often store or process large amounts of state. State consists of data, such as a database, that is frequently updated. Contrast this with a corpus, which is relatively static or is updated only periodically when a new edition is published. For example, a system that searches the U.S. Library of Congress may receive a new corpus each week. By comparison, an email system is in constant churn with new data arriving constantly, current data being updated (email messages being marked as "read" or moved between folders), and data being deleted.

Distributed computing systems have many ways to deal with state. However, they all involve some kind of replication and sharding, which brings about problems of consistency, availability, and partitioning.

The easiest way to store state is to put it on one machine, as depicted in Figure 10.4. Unfortunately, that method reaches its limit quite quickly: an individual machine can store only a limited amount of state and if the one machine dies we lose access to 100 percent of the state. The machine has only a certain amount of processing power, which means the number of simultaneous reads and writes it can process is limited.

In distributed computing we store state by storing fractions or shards of the whole on individual machines. This way the amount of state we can store is limited only by the number of machines we can acquire. In addition, each shard is stored on multiple machines; thus a single machine failure does not lose access to any state. Each replica can process a certain number of queries per second, so we can design the system to process any number of simultaneous read and write requests by increasing the number of replicas. This is illustrated in Figure 10.5, where N QPS are received and distributed among three shards, each replicated three ways. As a result, on average one ninth of all queries reach a particular replica server.

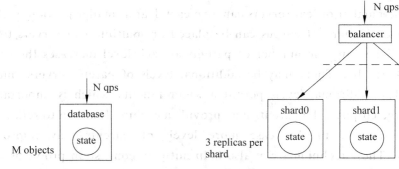

Figure 10.4　State kept in one location; not distributed computing.

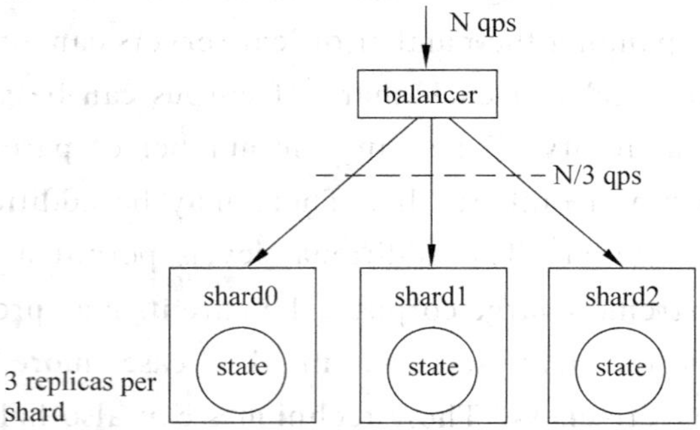

Figure 10.5　This distributed state is sharded and replicated.

Writes or requests that update state require all replicas to be updated. While this update process is happening, it is possible that some clients will read from stale replicas that have not yet been updated. Figure 10.6 illustrates how a write can be confounded by reads to an out-of-date cache. This will be discussed further in the next section.

In the most simple pattern, a root server receives requests to store or retrieve state. It determines which shard contains that part of the state and forwards the

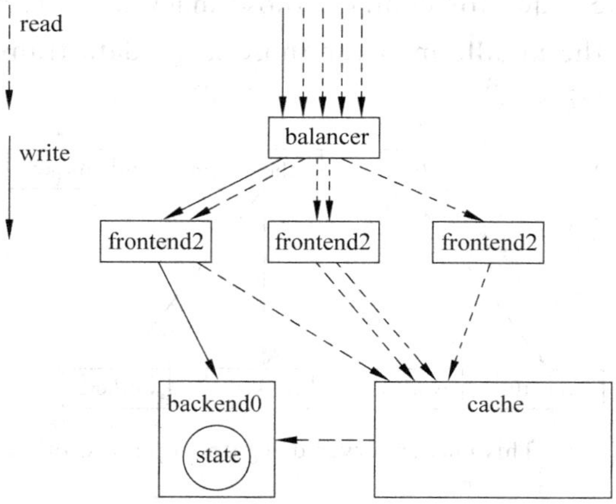

Figure 10.6 State updates using cached data lead to an inconsistent view.

request to the appropriate leaf server. The reply then flows up the tree. This looks similar to the server tree pattern described in the previous section but there are two differences. First, queries go to a single leaf instead of all leaves. Second, requests can be update (write) requests, not just read requests. Updates are more complex when a shard is stored on many replicas. When one shard is updated, all of the replicas must be updated, too. This may be done by having the root update all leaves or by the leaves communicating updates among themselves.

A variation of that pattern is more appropriate when large amounts of data are being transferred. In this case, the root replies with instructions on how to get the data rather than the data itself. The requestor then requests the data from the source directly.

For example, imagine a distributed file system with petabytes of data spread out over thousands of machines. Each file is split into gigabyte-sized chunks. Each chunk is stored on multiple machines for redundancy. This scheme also permits the creation of files larger than those that would fit on one machine. A master server tracks the list of files and identifies where their chunks are. If you are familiar with the UNIX file system, the master can be thought of as storing the inodes, or per-file lists of data blocks, and the other machine as storing the actual blocks of data. File system operations go through a master server that uses the inode-like information to determine which machines to involve in the operation.

Imagine that a large read request comes in. The master determines that the file has a few terabytes stored on one machine and a few terabytes stored on another machine. It could request the data from each machine and relay it to the system that made the request, but the master would quickly become overloaded while receiving and relaying huge chunks of data. Instead, it replies with a list of which machines have the data, and the requestor contacts those machines directly for the data. This way the master is not the middle man for those large data transfers. This situation is illustrated in Figure 10.7.

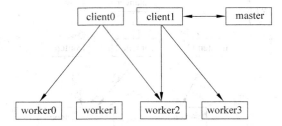

Figure 10.7 This master server delegates replies to other servers.

10.5 The CAP Principle

CAP stands for consistency, availability, and partition resistance. The CAP Principle states that it is not possible to build a distributed system that guarantees consistency, availability, and resistance to partitioning. Any one or two can be achieved but not all three simultaneously. When using such systems you must be aware of which are guaranteed.

10.5.1 Consistency

Consistency means that all nodes see the same data at the same time. If there are multiple replicas and there is an update being processed, all users see the update go live at the same time even if they are reading from different replicas. Systems that do not guarantee consistency may provide **eventual consistency**. For example, they may guarantee that any update will propagate to all replicas in a certain amount of time. Until that deadline is reached, some queries may receive the new data while others will receive older, out-of-date answers.

Perfect consistency is not always important. Imagine a social network that awards reputation points to users for positive actions. Your reputation point total is displayed anywhere your name is shown. The reputation database is replicated in the United States, Europe, and Asia. A user in Europe is awarded points and that change might take minutes to propagate to the United States and Asia replicas. This may be sufficient for such a system because an absolutely accurate reputation score is not essential. If a user in the United States and one in Asia were talking on the phone as one was awarded points, the other user would see the update seconds later and that would be okay. If the update took minutes due to network congestion or hours due to a network outage, the delay would still not be a terrible thing.

Now imagine a banking application built on this system. A person in the United States and another in Europe could coordinate their actions to withdraw money from the same account at the same time. The ATM that each person uses would query its nearest database replica, which would claim the money is available and may be withdrawn. If the updates propagated slowly enough, both people would have the cash before the bank realized the money was already gone.

10.5.2 Availability

Availability is a guarantee that every request receives a response about whether it was successful or failed. In other words, it means that the system is up. For example, using many replicas to store data such that clients always have access to at least one working replica guarantees availability.

The CAP Principle states that availability also guarantees that the system is able to report failure. For example, a system may detect that it is overloaded and reply to requests with an error code that means "try again later." Being told this immediately is more favorable than having to wait minutes or hours before one gives up.

10.5.3 Partition Tolerance

Partition tolerance means the system continues to operate despite arbitrary message loss or failure of part of the system. The simplest example of partition tolerance is when the system continues to operate even if the machines involved in providing the service lose the ability to communicate with each other due to a network link going down (see Figure 10.8).

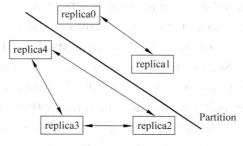

Figure 10.8　Nodes partitioned from each other.

Returning to our example of replicas, if the system is read-only it is easy to make the system partition tolerant, as the replicas do not need to communicate with each other. But consider the example of replicas containing state that is updated on one replica first, then copied to other replicas. If the replicas are unable to communicate with each other, the system fails to be able to guarantee updates will propagate within a certain amount of time, thus becoming a failed system.

Now consider a situation where two servers cooperate in a master-slave relationship. Both maintain a complete copy of the state and the slave takes over the master's role if the master fails, which is determined by a loss of heartbeat—that is, a periodic health check between two servers often done via a dedicated network. If the heartbeat network between the two is partitioned, the slave will promote itself to being the master, not knowing that the original master is up but unable to communicate on the heartbeat network. At this point there are two masters and the system breaks. This situation is called **split brain**.

Some special cases of partitioning exist. Packet loss is considered a temporary partitioning of the system as it applies to the CAP Principle. Another special case is the complete network outage. Even the most partition-tolerant system is unable to work in that situation.

The CAP Principle says that any one or two of the attributes are achievable in combination, but not all three. In 2002, Gilbert and Lynch published a formal proof of the original conjecture, rendering it a theorem. One can think of this as the third attribute being sacrificed to achieve the other two.

The CAP Principle is illustrated by the triangle in Figure 10.9. Traditional relational databases like Oracle, MySQL, and PostgreSQL are consistent and available (CA). They use transactions and other database techniques to assure that updates are atomic; they propagate completely or not at all. Thus they guarantee all users will see the same state at the same time. Newer storage systems such as Hbase, Redis, and Bigtable focus on consistency and partition tolerance (CP). When partitioned, they become read-only or refuse to respond to any requests rather than be inconsistent and permit some users to see old data while others see fresh data. Finally, systems such as Cassandra, Risk, and Dynamo focus on availability and partition tolerance (AP). They emphasize always being able to serve requests even if it means some clients receive outdated results. Such systems are often used in globally distributed networks where each replica talks to the others by less reliable media such as the Internet.

SQL and other relational databases use the term **ACID** to describe their side of the CAP triangle. ACID stands for Atomicity (transactions are "all or nothing"), Consistency (after each transaction the database is in a valid state), Isolation (concurrent transactions give the same results as if they were executed serially), and Durability (a committed transaction's data will not be lost in the event of a crash or other problem). Databases that provide weaker consistency models often refer to themselves as NoSQL and describe themselves as **BASE**: Basically Available Soft-state

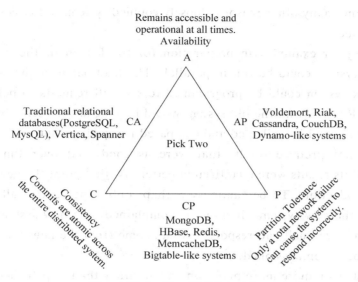

Figure 10.9 The CAP Principle.

services with Eventual consistency.

10.6 Loosely Coupled Systems

Distributed systems are expected to be highly available, to last a long time, and to evolve and change without disruption. Entire subsystems are often replaced while the system is up and running.

To achieve this a distributed system uses **abstraction** to build a loosely coupled system. Abstraction means that each component provides an interface that is defined in a way that hides the implementation details. The system is loosely coupled if each component has little or no knowledge of the internals of the other components. As a result a subsystem can be replaced by one that provides the same abstract interface even if its implementation is completely different.

Take, for example, a spell check service. A good level of abstraction would be to take in text and return a description of which words are misspelled and a list of possible corrections for each one. A bad level of abstraction would simply provide access to a lexicon of words that the frontends could query for similar words. The reason the latter is not a good abstraction is that if an entirely new way to check spelling was invented, every frontend using the spell check service would need to be rewritten. Suppose this new version does not rely on a lexicon but instead applies an artificial intelligence technique called machine learning. With the good abstraction, no frontend would need to change; it would simply send the same kind of request to the new server. Users of the bad abstraction would not be so lucky.

For this and many other reasons, loosely coupled systems are easier to evolve and change over time.

Continuing our example, in preparation for the launch of the new spell check service both versions could be run in parallel. The load balancer that sits in front of the spell check system could be programmed to send all requests to both the old and new systems. Results from the old system would be sent to the users, but results from the new system would be collected and compared for quality control. At first the new system might not produce results that were as good, but over time it would be enhanced until its results were quantifiably better. At that point the new system would be put into production. To be cautious, perhaps only 1 percent of all queries would come through the new system—if no users complained, the new system would take a larger fraction. Eventually all responses would come from the new system and the old system could be decommissioned.

Other systems require more precision and accuracy than a spell check system. For example, there may be requirements that the new system be bug-for-bug compatible with the old system before it can offer new functionality. That is, the new system must reproduce not only the features but also the bugs from the old system. In this case the ability to send requests to both systems and compare results becomes critical to the operational task of deploying it.

Case Study: Emulation before Improvements

When Tom was at Cibernet, he was involved in a project to replace an older system. Because it was a financial system, the new system had to prove it was bug-for-bug compatible before it could be deployed.

The old system was built on obsolete, pre-web technology and had become so complex and calcified that it was impossible to add new features. The new system was built on newer, better technology and, being a cleaner design, was more easily able to accommodate new functionality. The systems were run in parallel and results were compared.

At that point engineers found a bug in the old system. Currency conversion was being done in a way that was non-standard and the results were slightly off. To make the results between the two systems comparable, the developers reverse-engineered the bug and emulated it in the new system.

Now the results in the old and new systems matched down to the penny. With the company having gained confidence in the new system's ability to be bug-for-bug compatible, it was activated as the primary system and the old system was disabled.

At this point, new features and improvements could be made to the system. The first improvement, unsurprisingly, was to remove the code that emulated the currency conversion bug.

10.7 Speed

So far we have elaborated on many of the considerations involved in designing large distributed systems. For web and other interactive services, one item may be the most important: speed. It takes time to get information, store information, compute and transform information, and transmit information. Nothing happens instantly.

An interactive system requires fast response times. Users tend to perceive anything faster than 200 ms to be instant. They also prefer fast over slow. Studies have documented sharp drops in revenue when delays as little as 50 ms were artificially added to web sites. Time is also important in batch and non-interactive systems where the total throughput must meet or exceed the incoming flow of work.

The general strategy for designing a system that is performant is to design a system using our best estimates of how quickly it will be able to process a request and then to build prototypes to test our assumptions. If we are wrong, we go back to step one; at least the next iteration will be informed by what we have learned. As we build the system, we are able to remeasure and adjust the design if we discover our estimates and prototypes have not guided us as well as we had hoped.

At the start of the design process we often create many designs, estimate how fast each will be, and eliminate the ones that are not fast enough. We do not automatically select the fastest design. The fastest design may be considerably more expensive than one that is sufficient.

How do we determine if a design is worth pursuing? Building a prototype is very time consuming. Much can be deduced with some simple estimating exercises. Pick a few common transactions and break them down into smaller steps, and then estimate how long each step will take.

Two of the biggest consumers of time are disk access and network delays.

Disk accesses are slow because they involve mechanical operations. To read a block of data from a disk requires the read arm to move to the right track; the platter must then spin until the desired block is under the read head. This process typically takes 10 ms. Compare this to reading the same amount of information from RAM, which takes 0.002 ms, which is 5,000 times faster. The arm and platters (known as a **spindle**) can process only one request at a time. However, once the head is on the right track, it can read many sequential blocks. Therefore reading two blocks is often nearly as fast as reading one block if the two blocks are adjacent. Solid-state drives (SSDs) do not have mechanical spinning platters and are much faster, though more expensive.

Network access is slow because it is limited by the speed of light. It takes approximately 75 ms for a packet to get from California to the Netherlands. About half of that journey time is due to the speed of light. Additional delays may be attributable to processing time on each router, the electronics that convert from wired to fiber-optic communication and back, the time it takes to assemble and disassemble the packet on each end, and so on.

Two computers on the same network segment might seem as if they communicate instantly, but that is not really the case. Here the time scale is so small that other delays have a bigger factor. For example, when transmitting data over a local network, the first byte arrives quickly but the program receiving the data usually does not process it until the entire packet is received.

In many systems computation takes little time compared to the delays from network and disk operation. As a result you can often estimate how long a transaction will take if you simply know the distance from the user to the datacenter and the number of disk seeks required. Your estimate will often be good enough to throw away obviously bad designs.

To illustrate this, imagine you are building an email system that needs to be able to retrieve a message from the message storage system and display it within 300 ms. We will use the time approximations listed in Figure 10.10 to help us engineer the solution.

Jeff Dean, a Google Fellow, has popularized this chart of common numbers to aid in architectural and scaling decisions. As you can see, there are many orders of magnitude difference between certain options. These numbers improve every year. Updates can be found online.

Action	Typical Time	
L1 cache reference	0.5 ns	
Branch mispredict	5 ns	
L2 cache reference	7 ns	
Mutex lock/unlock	100 ns	
Main memory reference	100 ns	
Compress 1K bytes with Zippy	10,000 ns	(0.01 ms)
Send 2K bytes over 1 Gbps network	20,000 ns	(0.02 ms)
Read 1 MB sequentially from memory	250,000 ns	(0.25 ms)
Round trip within same datacenter	500,000 ns	(0.5 ms)
Read 1 MB from SSD	1,000,000 ns	(3 ms)
Disk seek	10,000,000 ns	(10 ms)
Read 1 MB sequentially from network	10,000,000 ns	(10 ms)
Read 1 MB sequentially from disk	30,000,000 ns	(30 ms)
Send packet from California to Netherlands to California	150,000,000 ns	(150 ms)

Figure 10.10　Numbers every engineer should know.

First we follow the transaction from beginning to end. The request comes from a web browser that may be on another continent. The request must be authenticated, the database index is consulted to determine where to get the message text, the message text is retrieved, and finally the response is formatted and transmitted back to the user.

Now let's budget for the items we can't control. To send a packet between California and Europe typically takes 75 ms, and until physics lets us change the speed of light that won't change. Our 300 ms budget is reduced by 150 ms since we have to account for not only the time it takes for the request to be transmitted but also the reply. That's half our budget consumed by something we don't control.

We talk with the team that operates our authentication system and they recommend budgeting 3 ms for authentication.

Formatting the data takes very little time—less than the slop in our other estimates—so we can ignore it.

This leaves 147 ms for the message to be retrieved from storage. If a typical index lookup requires 3 disk seeks (10 ms each) and reads about 1 megabyte of information (30 ms), that is 60 ms. Reading the message itself might require 4 disk seeks and reading about 2 megabytes of information (100 ms). The total is 160 ms, which is more than our 147 ms remaining budget.

How Did We Know That?

How did we know that it will take 3 disk seeks to read the index? It requires knowledge of the inner workings of the UNIX file system: how files are looked up in a directory to find an inode and how inodes are used to look up the data blocks. This is why understanding the internals of the operating system you use is key to being able to design and operate distributed systems. The internals of UNIX and UNIX-like operating systems are well documented, thus giving them an advantage over other systems.

While disappointed that our design did not meet the design parameters, we are happy that disaster has been averted. Better to know now than to find out when it is too late.

It seems like 60 ms for an index lookup is a long time. We could improve that considerably. What if the index was held in RAM? Is this possible? Some quick calculations estimate that the lookup tree would have to be 3 levels deep to fan out to enough machines to span this much data. To go up and down the tree is 5 packets, or about 2.5 ms if they are all within the same datacenter. The new total (150 ms + 3 ms + 2.5 ms + 100 ms = 255.5 ms) is less than our total 300 ms budget.

We would repeat this process for other requests that are time sensitive. For example, we send email messages less frequently than we read them, so the time to send an email message may not be considered time critical. In contrast, deleting a message happens almost as often reading messages. We might repeat this calculation for a few deletion methods to compare their efficiency.

One design might contact the server and delete the message from the storage system and the index. Another design might have the storage system simply mark the message as deleted in the index. This would be considerably faster but would require a new element that would reap messages marked for deletion and occasionally compact the index, removing any items marked as deleted.

Even faster response time can be achieved with an asynchronous design. That means the client sends requests to the server and quickly returns control to the user without waiting for the request to complete. The user perceives this system as faster even though the actual work is lagging. Asynchronous designs are more complex to implement. The server might queue the request rather than actually performing the action. Another process reads requests from the queue and performs them in the background. Alternatively, the client could simply send the request and check for the reply later, or allocate a thread or subprocess to wait for the reply.

All of these designs are viable but each offers different speed and complexity of implementation. With speed and cost estimates, backed by prototypes, the business decision of which to implement can be made.

10.8 Summary

Distributed computing is different from traditional computing in many ways. The scale is larger; there are many machines, each doing specialized tasks. Services are replicated to increase capacity. Hardware failure is not treated as an emergency or exception but as an expected part of the system. Thus the system works around failure.

Large systems are built through composition of smaller parts. We discussed three ways this composition is typically done: load balancer for many backend replicas, frontend with many different backends, and a server tree.

The load balancer divides traffic among many duplicate systems. The front-end with many different backends uses different backends in parallel, with each performing different processes. The server tree uses a tree configuration, with each tree level serving a different purpose.

Maintaining state in a distributed system is complex, whether it is a large database of constantly updated information or a few key bits to which many systems need constant access. The CAP Principle states that it is not possible to build a distributed system that guarantees consistency, availability, and resistance to partitioning simultaneously. At most two of the three can be achieved.

Systems are expected to evolve over time. To make this easier, the components are loosely coupled. Each embodies an abstraction of the service it provides, such that the internals can be replaced or improved without changing the abstraction. Thus, dependencies on the service do not need to change other than to benefit from new features.

Designing distributed systems requires an understanding of the time it takes various operations to run so that time-sensitive processes can be designed to meet their latency budget.

10.9 New Words and Expressions

eponymous	*adj.* 使得名的,齐名的
neologism	*n.* 新词;新义
introspection	*n.* 内省;反省
load balancer	负载平衡器
replicas	*n.* 复制品;复型
CNN (Cable News Network)	*abbr.* 美国有线电视新闻网络
corpus	*n.* 语料库;文集
encyclopedia	*n.* 百科全书(同 encyclopaedia)
individuals	*n.* 个人;个体
Library of Congress	*n.* 美国国会图书馆
even less	何况,更不用说;更何况(= still less)
redundant server	冗余服务器
simultaneous	*adj.* 同时的;同时发生的
retrieve	*n.* 检索;恢复
eventual consistency	最终一致性
lexicon	*n.* 词典;辞典
emulation	*n.* 仿真
calcified	*adj.* 钙化的;石灰化的
authentication	*n.* 证明;鉴定
emergency	*n.* 紧急情况;突发事件

10.10 Questions

Single Choice Questions

1. How many ms do users consider a web site sluggish if replies do not come back in or less? ( )
  A. 10   B. 20   C. 100   D. 200

2. Which of the following statements is true? ( )
  A. The more simple the system, the more difficult it is to have an accurate mental model

B. The more complex the system, the more difficult it is to have an accurate mental model

C. The more complex the system, the more easy it is to have an accurate mental model

D. The more difficult the system, the more complex it is to have an accurate mental model

3. Which of the following statements is true? (   )

   A. The frontend receives the request. It relays the query to many frontend servers

   B. The frontend receives the request. It relays the query to many backend servers

   C. The backend receives the request. It relays the query to many frontend servers

   D. The backend receives the request. It relays the query to many backend servers

4. What is the primary benefit of the pattern of a server tree? (   )

   A. It permits parallel searching of a large corpus
   B. It permits parallel searching of a small corpus
   C. It permits serial searching of a large corpus
   D. It permits serial searching of a small corpus

5. Which of the following does the CAP stand for? (   )

   A. Consistency   B. Availability
   C. Partition resistance   D. All of the above

6. Which of the following statements is true? (   )

   A. If the heartbeat network between the two is partitioned, the slave will promote itself to being the master

   B. If the heartbeat network between the two is partitioned, the master will promote itself to being the slave

   C. If the heartbeat network between the two is partitioned, the slave will still be the slave

   D. If the heartbeat network between the two is partitioned, the master will still be the master

7. What kind of newer storage systems focus on consistency and partition tolerance (CP)? (   )

   A. Hbase   B. Redis   C. Bigtable   D. All of the above

8. Which of the following item may be the most important for web and other interactive services? (   )

   A. Cost   B. Size   C. Speed   D. Storage

9. What are the two of the biggest consumers of time? (　　)
   A. Disk access and network access
   B. Disk access and network delays
   C. Disk delays and network access
   D. Disk delays and network delays
10. Which of the following statements is true? (　　)
    A. We discussed one way this composition is typically done: load balancer for many backend replicas
    B. We discussed two ways this composition is typically done: load balancer for many backend replicas and frontend with many different backends
    C. We discussed three ways this composition is typically done: load balancer for many backend replicas, frontend with many different backends, and a server tree
    D. None of the above

10.11　Problems

After reading this chapter and completing the exercises, you will be able to do the following:

1. What is distributed computing?
2. Describe the three major composition patterns in distributed computing.
3. What are the three patterns discussed for storing state?
4. Sometimes a master server does not reply with an answer but instead replies with where the answer can be found. What are the benefits of this method?
5. Section 10.4 describes a distributed file system, including an example of how reading terabytes of data would work. How would writing terabytes of data work?
6. Explain the CAP Principle. (If you think the CAP Principle is awesome, read "The Part-Time Parliament" (Lamport & Marzullo 1998) and "Paxos Made Simple" (Lamport 2001).)
7. What does it mean when a system is loosely coupled? What is the advantage of these systems?
8. Give examples of loosely and tightly coupled systems you have experience with. What makes them loosely or tightly coupled?
9. How do we estimate how fast a system will be able to process a request such as retrieving an email message?
10. In Section 10.7 three design ideas are presented for how to process email deletion requests. Estimate how long the request will take for deleting an email message for each of the three designs. First outline the steps each would take, then break each one into individual operations until estimates can be created.

Unit 11　Big Data—Big Data Computing

11.1　Introduction

The growing demands of today's applications such as, business, government, defense, surveillance, agencies, aerospace, research, development and entertainment sector has generated multitude of data. Intensive competition and customer's information have pushed the new technique to innovate. The idea of big data computing is concept for data sets so large and complex, as shown in Figure 11.1. Thus, the conventional data processing technique is inadequate to process the data set. Currently, the big data and cloud computing are attracted more attentions. The cloud computing is the solution for the incapable storage with massive data. The data is usually from social media sites such as Facebook, Twitter and Youtube. The challenges are analysis, capture, data curation, search, sharing, storage, transfer, visualization, and information privacy. The most often terms are simply used predictive analytics or other certain advanced methods to extract value from database, and compressed to a particular small size of data. The accuracy is another concern about more confident decision making, and the better decision is meaning more operational efficiency. The radically new approaches to research data modelling are needed to covering **Data Modelling Challenges**. Current data models (relational model) and management systems (relational database management systems) were developed by the database research community for business/commercial data applications. Research data has completely different characteristics and thus the current database technology is unable to handle it effectively. There is a need for data models and

Figure 11.1　Big data's biggest vision

query languages that:
- More closely match the data representation needs of the several scientific disciplines;
- Describe discipline-specific aspects (metadata models);
- Represent and query data provenance information;
- Represent and query data contextual information;
- Represent and manage data uncertainty;
- Represent and query data quality information.

Data management is a clear need for extremely large data processing to cover Data Management Challenges. This is especially true in the area of scientific data management where, in the near future, we expect data inputs in the order of multiple Petabytes. However, in the case shown in Figure 11.2, current data management technology is not suitable for such data sizes. In the light of such new database applications, we need to rethink some of the strict requirements adopted by database systems in the past. For instance, database management systems (DBMS) see database queries as contracts carved in stone that require the DBMS to produce a complete and correct answer, regardless of the time and resources required. While this behavior is crucial in business data management, it is counterproductive in scientific data management. With the explorative nature of scientific discovery, scientists cannot be expected to instantly phrase a crisp query that yields the desired (but a priori unknown) result, or to wait days to get a multi-megabyte answer that does not reveal what they were looking for. Instead, the DBMS could provide a fast and cheap approximation that is neither complete nor correct, but indicative of the complete answer. In this way, the user gets a 'feel' for the data that helps him/her to advance his/her exploration using a refined query. The challenges faced include catching the user's intention and providing the users with suggestions and guidelines to refine their queries in order to quickly converge to the desired results, but also call for novel database architectures and algorithms that are designed with the intent to produce fast and cheap indicative answers rather than complete and correct answers. Currently, the available data tools for most scientific disciplines are inadequate to cover **Data Tools Challenges.** It is essential to build better tools in order to improve the productivity of scientists. There is a need for better computational tools to visualize, analyze, and catalog the available enormous research datasets in order to enable data-driven research. Scientists need advanced tools that enable them to follow new paths, try new techniques, build new models and test them in new ways that facilitate innovative multidisciplinary/interdisciplinary activities and support the whole research cycle.

Figure 11.2　Big data reports from most famous journals

11.2　Apache Hadoop Data modelling

Apache Hadoop is an open source software framework for storage and large scale processing of data-sets on clusters of commodity hardware. Hadoop is an Apache top-level project being built and used by a global community of contributors and users. It is licensed under the Apache License 2.0.

The Apache Hadoop framework is composed of the following modules, as shown in Figure 11.3.

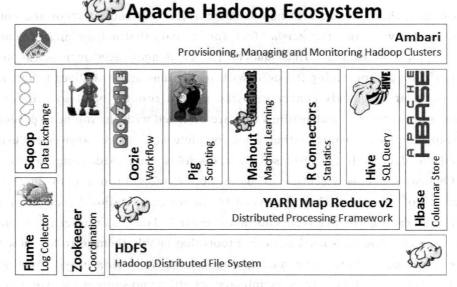

Figure 11.3　Apache Hadoop ecosystem structure

1. Hadoop Common: contains libraries and utilities needed by other Hadoop modules.

2. Hadoop Distributed File System (HDFS): a distributed file-system that stores data on the commodity machines, providing very high aggregate bandwidth across the cluster.

3. Hadoop YARN: a resource-management platform responsible for managing compute resources in clusters and using them for scheduling of users' applications.

4. Hadoop MapReduce: a programming model for large scale data processing.

11.3　NoSQL Big Data systems

NoSQL Big Data systems are designed to take advantage of new cloud computing architectures that have emerged over the past decade to allow massive computations to be run inexpensively and efficiently. This makes operational big data workloads much easier to manage, cheaper, and faster to implement. Some NoSQL systems can provide insights into patterns and trends based on real-time data with minimal coding and without the need for data scientists and additional infrastructure.

Types of NoSQL datastores

11.3.1　Key Value stores

Examples: Tokyo Cabinet/Tyrant, Redis, Voldemort, Oracle BDB

Typical applications: Content caching Strengths: Fast lookups Weaknesses: Stored data has no schema.

Example application: You are writing forum software where you have a home profile page that gives the user's statistics (messages posted, etc) and the last ten messages by them. The page reads from a key that is based on the user's id and retrieves a string of JSON that represents all the relevant information. A background process recalculates the information every 15 minutes and writes to the store independently.

11.3.2　Document databases

Examples: CouchDB, MongoDb Typical applications: Web applications Strengths: Tolerant of incomplete data Weaknesses: Query performance, no standard query syntax

Example application: You are creating software that creates profiles of refugee children with the aim of reuniting them with their families. The details you need to record for each child vary tremendously with circumstances of the event and they are built up piecemeal, for example a young child may know their first name and you can take a picture of them but they may not know their parent's first names. Later a local may claim to recognize the child and provide you with additional information that you definitely want to record but until you can verify the information you have to treat it skeptically.

11.3.3 Graph databases

Examples: Neo4J, InfoGrid, Infinite Graph Typical applications: Social networking, Recommendations Strengths: Graph algorithms e. g. shortest path, connectedness, n degree relationships, etc. Weaknesses: Has to traverse the entire graph to achieve a definitive answer. Not easy to cluster.

Example application: Any application that requires social networking is best suited to a graph database. These same principles can be extended to any application where you need to understand what people are doing, buying or enjoying so that you can recommend further things for them to do, buy or like. Any time you need to answer the question along the lines of "What restaurants do the sisters of people who are over-40, enjoy skiing and have visited Kenya dislike?" a graph database will usually help.

11.3.4 XML databases

Examples: Exist, Oracle, MarkLogic Typical applications: Publishing Strengths: Mature search technologies, Schema validation Weaknesses: No real binary solution, easier to re-write documents than update them

Example application: A publishing company that uses bespoke XML formats to produce web, print and eBook versions of their articles. Editors need to quickly search either text or semantic sections of the markup (e.g. articles whose summary contains diabetes, where the author's institution is Liverpool University and Stephen was a revising editor at some point in the document history). They store the XML of finished articles in the XML database and wrap it in a readable-URL web service for the document production systems. Workflow metadata (which stage a manuscript is in) is held in a separate RDBMS. When system-wide changes are required, XQuery updates bulk update all the documents to match the new format.

11.3.5 Distributed Peer Stores

Examples: Cassandra, HBase, Riak Typical applications: Distributed file systems

Strengths: Fast lookups, good distributed storage of data Weaknesses: Very low-level API Example application: You have a news site where any piece of content: articles, comments, author profiles, can be voted on and an optional comment supplied on the vote. You create one store per user and one store per piece of content, using a UUID as the key (generating one for each piece of content and user). The user's store holds every vote they have ever made while the content "bucket" contains a copy of every vote that has been made on the piece of content. Overnight you run a batch job to identify content that users have voted on, you generate a list of content for each user that has high votes but which they have not voted on. You then push this list of recommended articles into the user's "bucket".

11.3.6 Object stores

Examples: Oracle Coherence, db4o, ObjectStore, GemStone, Polar Typical applications: Finance systems Strengths: Matches OO development paradigm, low-latency ACID, mature technology Weaknesses: Limited querying or batch-update options

Example application: A global trading company has a monoculture of development and wants to have trades done on desks in Japan and New York pass through a risk checking process in London. An object representing the trade is pushed into the object store and the risk checker is listening to for appearance or modification of trade objects. When the object is replicated into the local European space the risk checker reads the Trade and assesses the risk. It then rewrites the object to indicate that the trade is approved and generates an actual trade fulfillment request. The trader's client is listening for changes to objects that contain the trader's id and updates the local detail of the trade in the client indicating to the trader that the trader has been approved. The trading system will consume the trade fulfillment and when the trade elapses or is fulfilled feeds back the information to the risk assessor.

11.4 Definitions of Data Management

11.4.1 Data management

This involves the collection and storage of data, plus its processing and delivery—whether traditional data, new big data, or both. Processing can be extensive, especially when data is repurposed for a use differing from that of its origin (as is common in business intelligence [BI], data warehousing [DW], and analytics). Data management is a broad practice that encompasses a number of data disciplines, including data warehousing, data integration, data quality, data governance, content management, event processing, database administration, and so on.

11.4.2 Big data management (BDM)

This is where data management disciplines, tools, and platforms (both old and new) are applied to the management of big data (in the base definition or the extended one). Traditional data and new big data can be quite different in terms of content, structure, and intended use, and each category has many variations within it. To accommodate this diversity, softwaresolutions for BDM tend to include multiple types of data management tools and platforms, as well as diverse user skills and practices.

11.5 The State of Big Data Management

A number of user organizations are actively managing big data today, as seen in survey results. An example is shown in Figure 11.4. However, do they manage big data with a dedicated BDM solution, as opposed to extending existing data management platforms?

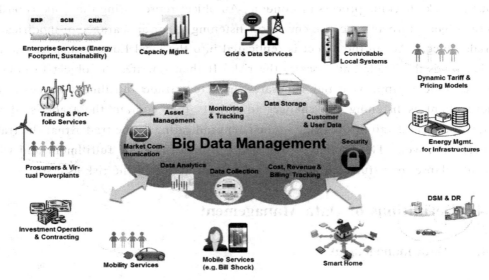

Figure 11.4 Big Data Management

Dedicated BDM solutions are quite rare, for the moment. Only 10% of respondents report having deployed a special solution for managing big data today. Most of these are very new (7%), whereas a few are relatively mature (3%). This is consistent with the 11% of respondents who already have a BDM solution in production.

In the short term, the number of deployed BDM solutions will double. Another 10% of respondents say they have a BDM solution in development as a committed project. This is consistent with the 10% who say they will deploy a dedicated BDM solution within six months.

Half of surveyed organizations plan to bring a BDM solution online within three years. In addition to the 10% over six months just noted, more solutions will come online in 12 months (20%), 24 months (19%), and 36 months (12%). If users' plans pan out, dedicated BDM solutions will jump from rare to mainstream within three years. But note that users' plans are by no means certain, because many projects are still in the prototyping or discussion stage (20% and 37%).

Few organizations don't need a special solution for managing big data. Just a quarter report no plans at present for such a solution (23%); even fewer say they'll never deploy a BDM solution (6%).

Strategies for Managing Big Data. Different organizations take different technology approaches to managing big data. On one hand, a "fork in the road" decision is whether to manage big data in existing data management platforms or to deploy one or more dedicated solutions just for managing big data. On the other hand, some organizations don't have or say they don't need a strategy for managing big data.

Half of organizations have a strategy for managing big data. This is true whether the strategy involves deploying new data management systems specifically for big data (20%) or extending existing systems to accommodate big data (31%). One survey respondent selected "other" and added the comment: "Our big data strategy is a core competency for our business."

The other half doesn't have a strategy, for various reasons. Some don't have a strategy because they're not committed to big data (15%). "The business value is questionable," said one respondent. Others lack a strategy for managing big data, as yet, even though they know they need one (30%). "Once our POC completes, strategy can be defined."

A lack of maturity can prevent a strategy from coalescing. One survey respondent added the comment: "We don't know enough yet to determine a strategy." Another commented: "Our data management is in a nascent stage. It needs to mature before a strategy becomes clear."

As with many strategies, hybrids can be useful. According to one respondent: "We'll use a blend of extending existing platforms and deploying new ones in a hybrid mode." Another echoed that strategy, but turned it into an evolutionary process: "We'll extend existing systems now, and add new and better systems later."

11.6 Big Data Tools

Hadoop is a popular tool for organizing the racks and racks of servers, and NoSQL databases are popular tools for storing data on these racks. These mechanism can be much more powerful than the old single machine, but they are far from being as polished as the old database servers. Although SQL may be complicated, writing the JOIN query for the SQL databases was often much simpler than gathering information from dozens of machines and compiling it into one coherent answer. Hadoop jobs are written in Java, and that requires another level of sophistication. The tools for tackling big data are just beginning to package this distributed computing power in a way that's a bit easier to use.

11.6.1 Big data tools: Jaspersoft BI Suite

The Jaspersoft package is one of the open source leaders for producing reports from database columns. The software is well-polished and already installed in many businesses turning SQL tables into PDFs that everyone can scrutinize at meetings.

The company is jumping on the big data train, and this means adding a software layer to connect its report generating software to the places where big data gets stored. The JasperReports Server now offers software to suck up data from many of the major storage platforms, including MongoDB, Cassandra, Redis, Riak, CouchDB, and Neo4j. Hadoop is also well-represented, with JasperReports providing a Hive connector to reach inside of HBase.

This effort feels like it is still starting up—many pages of the documentation wiki are blank, and the tools are not fully integrated. The visual query designer, for instance, doesn't work yet with Cassandra's CQL. You get to type these queries out by hand.

Once you get the data from these sources, Jaspersoft's server will boil it down to interactive tables and graphs. The reports can be quite sophisticated interactive tools that let you drill down into various corners. You can ask for more and more details if you need them.

This is a well-developed corner of the software world, and Jaspersoft is expanding by making it easier to use these sophisticated reports with newer sources of data. Jaspersoft isn't offering particularly new ways to look at the data, just more sophisticated ways to access data stored in new locations. I found this surprisingly useful. The aggregation of my data was enough to make basic sense of who was going to the website and when they were going there.

11.6.2　Big data tools: Pentaho Business Analytics

Pentaho is another software platform that began as a report generating engine; it is, like JasperSoft, branching into big data by making it easier to absorb information from the new sources. You can hook up Pentaho's tool to many of the most popular NoSQL databases such as MongoDB and Cassandra. Once the databases are connected, you can drag and drop the columns into views and reports as if the information came from SQL databases.

I found the classic sorting and sifting tables to be extremely useful for understanding just who was spending the most amount of time at my website. Simply sorting by IP address in the log files revealed what the heavy users were doing.

Pentaho also provides software for drawing HDFS file data and HBase data from Hadoop clusters. One of the more intriguing tools is the graphical programming interface known as either Kettle or Pentaho Data Integration. It has a bunch of built-in modules that you can drag and drop onto a picture, then connect them. Pentaho has thoroughly integrated Hadoop and the other sources into this, so you can write your code and send it out to execute on the cluster.

11.6.3　Big data tools: Karmasphere Studio and Analyst

Many of the big data tools did not begin life as reporting tools. Karmasphere Studio, for instance, is a set of plug-ins built on top of Eclipse. It's a specialized IDE that makes it easier to create and run Hadoop jobs.

I had a rare feeling of joy when I started configuring a Hadoop job with this developer tool. There are a number of stages in the life of a Hadoop job, and Karmasphere's tools walk you through each step, showing the partial results along the way. I guess debuggers have always made it possible for us to peer into the mechanism as it does its work, but Karmasphere Studio does something a bit better: As you set up the workflow, the tools display the state of the test data at each step. You see what the temporary data will look like as it is cut apart, analyzed, then reduced.

Karmasphere also distributes a tool called Karmasphere Analyst, which is designed to simplify the process of plowing through all of the data in a Hadoop cluster. It comes with many useful building blocks for programming a good Hadoop job, like subroutines for uncompressing Zipped log files. Then it strings them together and parameterizes the Hive calls to produce a table of output for perusing.

11.6.4　Big data tools: Talend Open Studio

Talend also offers an Eclipse-based IDE for stringing together data processing jobs with Hadoop. Its tools are designed to help with data integration, data quality, and data management, all with subroutines tuned to these jobs.

Talend Studio allows you to build up your jobs by dragging and dropping little icons onto a canvas. If you want to get an RSS feed, Talend's component will fetch the RSS and add proxying if necessary. There are dozens of components for gathering information and dozens more for doing things like a "fuzzy match." Then you can output the results.

Stringing together blocks visually can be simple after you get a feel for what the components actually do and don't do. This was easier for me to figure out when I started looking at the source code being assembled behind the canvas. Talend lets you see this, and I think it's an ideal compromise. Visual programming may seem like a lofty goal, but I've found that the icons can never represent the mechanisms with enough detail to make it possible to understand what's going on. I need the source code.

Talend also maintains TalendForge, a collection of open source extensions that make it easier to work with the company's products. Most of the tools seem to be filters or libraries that link Talend's software to other major products such as Salesforce.com and SugarCRM. You can suck down information from these systems into your own projects, simplifying the integration.

11.6.5　Big data tools: Skytree Server

Not all of the tools are designed to make it easier to string together code with visual mechanisms. Skytree offers a bundle that performs many of the more sophisticated machine-learning algorithms. All it takes is typing the right command into a command line.

Skytree is more focused on the guts than the shiny GUI. Skytree Server is optimized to run a number of classic machine-learning algorithms on your data using an implementation the company claims can be 10,000 times faster than other packages. It can search through your data looking for clusters of mathematically similar items, then invert this to identify outliers that may be problems, opportunities, or both. The algorithms can be more precise than humans, and they can search through vast quantities of data looking for the entries that are a bit out of the ordinary. This may be fraud—or a particularly good customer who will spend and spend.

The free version of the software offers the same algorithms as the proprietary version, but it's limited to data sets of 100,000 rows. This should be sufficient to establish whether the software is a good match.

11.6.6　Big data tools: Tableau Desktop and Server

Tableau Desktop is a visualization tool that makes it easy to look at your data in new ways, then slice it up and look at it in a different way. You can even mix the data with other data and examine it in yet another light. The tool is optimized to give you all the columns for the data and let you mix them before stuffing it into one of the dozens of graphical templates provided.

Tableau Software started embracing Hadoop several versions ago, and now you can treat Hadoop "just like you would with any data connection." Tableau relies upon Hive to structure the queries, then tries its best to cache as much information in memory to allow the tool to be interactive. While many of the other reporting tools are built on a tradition of generating the reports offline, Tableau wants to offer an interactive mechanism so that you can slice and dice your data again and again. Caching helps deal with some of the latency of a Hadoop cluster.

The software is well-polished and aesthetically pleasing. I often found myself reslicing the data just to see it in yet another graph, even though there wasn't much new to be learned by switching from a pie chart to a bar graph and beyond. The software team clearly includes a number of people with some artistic talent.

11.6.7　Big data tools: Splunk

Splunk is a bit different from the other options. It's not exactly a report-generating tool or a collection of AI routines, although it accomplishes much of that along the way. It creates an index of your data as if your data were a book or a block of text. Yes, databases also build indices, but Splunk's approach is much closer to a text search process.

This indexing is surprisingly flexible. Splunk comes already tuned to my particular application, making sense of log files, and it sucked them right up. It's also sold in a number of different solution packages, including one for monitoring a Microsoft Exchange server and another for detecting Web attacks. The index helps correlate the data in these and several other common server-side scenarios.

Splunk will take text strings and search around in the index. You might type in the URLs of important articles or the IP address. Splunk finds them and packages them into a timeline built around the time stamps it discovers in the data. All other fields are correlated, and you can click around to drill deeper and deeper into the data set. While this is a simple process, it's quite powerful if you're looking for the right kind of needle in your data feed. If you know the right text string, Splunk will help you track it. Log files are a great application for it.

A new Splunk tool called Shep, currently in private beta, promises bidirectional integration between Hadoop and Splunk, allowing you to exchange data between the systems and query Splunk data from Hadoop.

11.7 New Words and Expressions

approximation	n. 近似法；接近
algorithm	n. 算法，运算法则
indicative	adj. 象征的；表示⋯的
facilitate	vt. 促进；帮助
innovative multidisciplinary	创新的多学科
DFS(Distributed File System)	分布式文件系统
yarn	n. 奇谈，故事
JSON(JavaScript Object Notation)	abbr. 基于 JavaScript 语言的轻量级的数据交换格式
tremendously	adv. 非常地；可怕地；惊人地
circumstance	n. 情况；环境
skeptically	adv. 怀疑地
recommendation	n. 推荐；建议
UUID(Universally Unique Identifier)	abbr. 通用唯一标识符
Low-latency	低延迟
intriguing	adj. 有趣的；迷人的
aesthetically	adv. 审美地；美学观点上地

11.8 Questions

Single Choice Questions

1. What kind of answer could the DBMS provide? （　　）

   A. A fast and cheap approximation that is neither complete nor correct

   B. A fast and cheap approximation that is both complete and correct

   C. A fast and cheap approximation that is complete but not correct

   D. A fast and cheap approximation that is correct but not complete

2. Which of the following modules is not composed of the Apache Hadoop

framework? ( )
    A. Hadoop Common   B. Hadoop File System
    C. Hadoop YARN   D. Hadoop MapReduce
3. How many percent of respondents report having deployed a special solution for managing big data today? ( )
    A. Only 10%   B. Only 20%   C. Only 30%   D. Only 40%
4. How long will it take that dedicated BDM solutions will jump from rare to mainstream if users' plans pan out? ( )
    A. within one year   B. within two years
    C. within three years   D. within four years
5. Which of the following statements is true? ( )
    A. Hadoop is a popular tool for organizing the racks and racks of servers, and for storing data on these racks
    B. NoSQL databases are popular tools for organizing the racks and racks of servers, and for storing data on these racks
    C. NoSQL databases are popular tools for organizing the racks and racks of servers, and Hadoop is a popular tool for storing data on these racks
    D. Hadoop is a popular tool for organizing the racks and racks of servers, and NoSQL databases are popular tools for storing data on these racks
6. Which of the following describes the function of the Jaspersoft package? ( )
    A. It is well-polished and already installed in many businesses turning NoSQL tables into PDFs that everyone can scrutinize at meetings
    B. It is well-polished and already installed in many businesses turning SQL tables into PDFs that everyone can scrutinize at meetings
    C. It is well-polished and already installed in many businesses turning PDFs tables into SQL that everyone can scrutinize at meetings
    D. It is well-polished and already installed in many businesses turning PDFs tables into NoSQL that everyone can scrutinize at meetings
7. Why do we say that Pentaho is like JasperSoft? ( )
    A. It is a software platform
    B. It began as a report generating engine
    C. It branched into big data by making it easier to absorb information from the new sources
    D. All of the above
8. Which of the following statements is true? ( )
    A. Pentaho also provides software for drawing HDFS file data and HBase data from Hadoop clusters
    B. HDFS also provides software for drawing Pentaho file data and HBase data

from Hadoop clusters
C. HBase also provides software for drawing Pentaho file data and HDFS data from Hadoop clusters
D. None of the above

9. What kind of the service does not belong to Talend? (   )
   A. Data integration   B. Data quality
   C. Data distribution   D. Data management

10. Which kind of the following channel for integration is allowed? (   )
    A. Hadoop promises bidirectional integration between Shep and Splunk
    B. Shep promises bidirectional integration between Hadoop and Splunk
    C. Splunk promises bidirectional integration between Hadoop and Shep
    D. None of the above

11.9 Problems

After reading this chapter and completing the exercises, you will be able to do the following:

1. Understand what is big data computing, and why we concern it.
2. Discuss the challenges of big data computing.
3. Describe current big data computing techniques.
4. Discuss modern tool development for big data computing.
5. Discuss the most significant applications of big data computing.

Unit 12 Cloud Computing—The Practice of Cloud System Administration: Operations in a Distributed World

This chapter starts with some operations management background, then discusses the operations service life cycle, and ends with a discussion of typical operations work strategies.

The rate at which organizations learn may soon become the only sustainable source of competitive advantage.

—Peter Senge

Now we discuss how to run distributed systems.

The work done to keep a system running is called **operations**. More specifically, operations is the work done to keep a system running in a way that meets or exceeds operating parameters specified by a service level agreement (SLA). Operations includes all aspects of a service's life cycle: from initial launch to the final decommissioning and everything in between.

Operational work tends to focus on availability, speed and performance, security, capacity planning, and software/hardware upgrades. The failure to do any of these well results in a system that is unreliable. If a service is slow, users will assume it is broken. If a system is insecure, outsiders can take it down. Without proper capacity planning, it will become overloaded and fail. Upgrades, done badly, result in downtime. If upgrades aren't done at all, bugs will go unfixed. Because all of these activities ultimately affect the reliability of the system, Google calls its operations team Site Reliability Engineering (SRE). Many companies have followed suit.

Operations is a team sport. Operations is not done by a single person but rather by a team of people working together. For that reason much of what we describe will be processes and policies that help you work as a team, not as a group of individuals. In some companies, processes seem to be bureaucratic mazes that slow things down. As we describe here—and more important, in our professional experience—good processes are exactly what makes it possible to run very large computing systems. In other words, process is what makes it possible for teams to do the right thing, again and again.

Terms to Know

Innovate: Doing (good) things we haven't done before.

Machine: A virtual or physical machine.

Oncall: Being available as first responder to an outage or alert.

Server: Software that provides a function or API. (Not a piece of hardware.)

Service: A user-visible system or product composed of one or more servers.

Soft launch: Launching a new service without publicly announcing it. This way traffic grows slowly as word of mouth spreads, which gives operations some cushion to fix problems or scale the system before too many people have seen it.

SRE: Site Reliability Engineer, the Google term for systems administrators who maintain live services.

Stakeholders: People and organizations that are seen as having an interest in a project's success.

This chapter starts with some operations management background, then discusses the operations service life cycle, and ends with a discussion of typical operations work strategies. All of these topics will be expanded upon in the chapters that follow.

12.1 Distributed Systems Operations

To understand distributed systems operations, one must first understand how it is different from typical enterprise IT. One must also understand the source of tension between operations and developers, and basic techniques for scaling operations.

12.1.1 SRE versus Traditional Enterprise IT

System administration is a continuum. On one end is a typical IT department, responsible for traditional desktop and client - server computing infrastructure, often called enterprise IT. On the other end is an SRE or similar team responsible for a distributed computing environment, often associated with web sites and other services. While this may be a broad generalization, it serves to illustrate some important differences.

SRE is different from an enterprise IT department because SREs tend to be focused on providing a single service or a well-defined set of services. A traditional enterprise IT department tends to have broad responsibility for desktop services, back-office services, and everything in between ("everything with a power plug"). SRE's customers tend to be the product management of the service while IT customers are the end users themselves. This means SRE efforts are focused on a few select business metrics rather than being pulled in many directions by users, each of whom has his or her own priorities.

Another difference is in the attitude toward uptime. SREs maintain services that have demanding, 24 × 7 uptime requirements. This creates a focus on preventing problems rather than reacting to outages, and on performing complex but non-intrusive maintenance procedures. IT tends to be granted flexibility with respect to scheduling downtime and has SLAs that focus on how quickly service can be restored in the event of an outage. In the SRE view, downtime is something to be avoided and service should not stop while services are undergoing maintenance.

SREs tend to manage services that are constantly changing due to new software releases and additions to capacity. IT tends to run services that are upgraded rarely. Often IT services are built by external contractors who go away once the system is stable.

SREs maintain systems that are constantly being scaled to handle more traffic and larger workloads. Latency, or how fast a particular request takes to process, is managed as well as overall throughput. Efficiency becomes a concern because a little waste per machine becomes a big waste when there are hundreds or thousands of machines. In IT, systems are often built for environments that expect a modest increase in workload per year. In this case a workable strategy is to build the system large enough to handle the projected workload for the next few years, when the system is expected to be replaced.

As a result of these requirements, systems in SRE tend to be bespoke systems, built on platforms that are home-grown or integrated from open source or other third-party components. They are not "off the shelf" or turn key systems. They are actively managed, while IT systems may be unchanged from their initial delivery state. Because of these differences, distributed computing services are best managed by a separate team, with separate management, with bespoke operational and management practices.

While there are many such differences, recently IT departments have begun to see a demand for uptime and scalability similar to that seen in SRE environments. Therefore the management techniques from distributed computing are rapidly being adopted in the enterprise.

12.1.2 Change versus Stability

There is a tension between the desire for stability and the desire for change. Operations teams tend to favor stability; developers desire change. Consider how each group is evaluated during end-of-the-year performance reviews. A developer is praised for writing code that makes it into production. Changes that result in a tangible difference to the service are rewarded above any other accomplishment. Therefore, developers want new releases pushed into production often. Operations, in contrast, is rewarded for achieving compliance with SLAs, most of which relate to uptime. Therefore stability is the priority.

A system starts at a baseline of stability. A change is then made. All changes have some kind of a destabilizing effect. Eventually the system becomes stable again, usually through some kind of intervention. This is called the **change-instability cycle**.

All software roll-outs affect stability. A change may introduce bugs, which are fixed through workarounds and new software releases. A release that introduces no new bugs still creates a destabilizing effect due to the process of shifting workloads away from machines about to be upgraded. Non-software changes also have a destabilizing effect. A network change may make the local network less stable while the change propagates throughout the network.

Because of the tension between the operational desire for stability and the developer desire for change, there must be mechanisms to reach a balance.

One strategy is to prioritize work that improves stability over work that adds new features. For example, bug fixes would have a higher priority than feature requests. With this approach, a major release introduces many new features, the next few releases focus on fixing bugs, and then a new major release starts the cycle over again. If engineering management is pressured to focus on new features and neglect bug fixes, the result is a system that slowly destabilizes until it spins out of control.

Another strategy is to align the goals of developers and operational staff. Both parties become responsible for SLA compliance as well as the velocity (rate of change) of the system. Both have a component of their annual review that is tied to SLA compliance and both have a portion tied to the on-time delivery of new features. Organizations that have been the most successful at aligning goals like this have restructured themselves so that developers and operations work as one team.

Another strategy is to budget time for stability improvements and time for new features. Software engineering organizations usually have a way to estimate the size of a software request or the amount of time it is expected to take to complete. Each new release has a certain size or time budget; within that budget a certain amount of stability-improvement work is allocated. Similarly, this allocation can be achieved by assigning dedicated people to stability-related code changes.

The budget can also be based on an SLA. A certain amount of instability is expected each month, which is considered a budget. Each roll-out uses some of the budget, as do instability-related bugs. Developers can maximize the number of roll-outs that can be done each month by dedicating effort to improve the code that causes this instability. This creates a positive feedback loop.

12.1.3 Defining SRE

The core practices of SRE were refined for more than 10 years at Google before being enumerated in public. In his keynote address at the first USENIX SREcon, Benjamin Treynor Sloss (2014), Vice President of Site Reliability Engineering at Google, listed them as follows:

Site Reliability Practices

1. Hire only coders.
2. Have an SLA for your service.
3. Measure and report performance against the SLA.
4. Use Error Budgets and gate launches on them.
5. Have a common staffing pool for SRE and Developers.
6. Have excess Ops work overflow to the Dev team.
7. Cap SRE operational load at 50 percent.
8. Share 5 percent of Ops work with the Dev team.
9. Oncall teams should have at least eight people at one location, or six people at each of multiple locations.
10. Aim for a maximum of two events per oncall shift.
11. Do a postmortem for every event.
12. Postmortems are blameless and focus on process and technology, not people.

The first principle for site reliability engineering is that SREs must be able to code. An SRE might not be a full-time software developer, but he or she should be able to solve nontrivial problems by writing code. When asked to do 30 iterations of a task, an SRE should do the first two, get bored, and automate the rest. An SRE must have enough software development experience to be able to communicate with developers on their level and have an appreciation for what developers do, and for what computers can and can't do.

When SREs and developers come from a common staffing pool, that means that projects are allocated a certain number of engineers; these engineers may be developers or SREs. The end result is that each SRE needed means one fewer developer in the team. Contrast this to the case at most companies where system administrators and developers are allocated from teams with separate budgets. Rationally a project wants to maximize the number of developers, since they write new features. The common staffing pool encourages the developers to create systems that can be operated efficiently so as to minimize the number of SREs needed.

Another way to encourage developers to write code that minimizes operational load is to require that excess operational work overflows to the developers. This practice discourages developers from taking shortcuts that create undue operational load. The developers would share any such burden. Likewise, by requiring developers to perform 5 percent of operational work, developers stay in tune with operational realities.

Within the SRE team, capping the operational load at 50 percent limits the amount of manual labor done. Manual labor has a lower return on investment than, for example, writing code to replace the need for such labor.

Many SRE practices relate to finding balance between the desire for change and the need for stability. The most important of these is the Google SRE practice called Error Budgets.

Central to the Error Budget is the SLA. All services must have an SLA, which specifies how reliable the system is going to be. The SLA becomes the standard by which all work is ultimately measured.

Any outage or other major SLA-related event should be followed by the creation of a written postmortem that includes details of what happened, along with analysis and suggestions for how to prevent such a situation in the future. This report is shared within the company so that the entire organization can learn from the experience. Postmortems focus on the process and the technology, not finding who to blame. The person who is oncall is responsible for responding to any SLA-related events and producing the postmortem report.

Oncall is not just a way to react to problems, but rather a way to reduce future problems. It must be done in a way that is not unsustainably stressful for those oncall, and it drives behaviors that encourage long-term fixes and problem prevention. Oncall teams are made up of at least eight members at one location, or six members at two locations. Teams of this size will be oncall often enough that their skills do not get stale, and their shifts can be short enough that each catches no more than two outage events. As a result, each member has enough time to follow through on each event, performing the required long-term solution.

Other companies have adopted the SRE job title for their system administrators who maintain live production services. Each company applies a different set of practices to the role. These are the practices that define SRE at Google and are core to its success.

12.1.4 Operations at Scale

Operations in distributed computing is operations at a large scale. Distributed computing involves hundreds and often thousands of computers working together. As a

Unit 12 Cloud Computing—The Practice of Cloud System Administration: Operations in a Distributed World

result, operations is different than traditional computing administration.

Manual processes do not scale. When tasks are manual, if there are twice as many tasks, there is twice as much human effort required. A system that is scaling to thousands of machines, servers, or processes, therefore, becomes untenable if a process involves manually manipulating things. In contrast, automation does scale. Code written once can be used thousands of times. Processes that involve many machines, processes, servers, or services should be automated. This idea applies to allocating machines, configuring operating systems, installing software, and watching for trouble. Automation is not a "nice to have" but a "must have."

When operations is automated, system administration is more like an assembly line than a craft. The job of the system administrator changes from being the person who does the work to the person who maintains the robotics of an assembly line. Mass production techniques become viable and we can borrow operational practices from manufacturing. For example, by collecting measurements from every stage of production, we can apply statistical analysis that helps us improve system throughput. Manufacturing techniques such as **continuous improvement** are the basis for the Three Ways of DevOps.

Three categories of things are not automated: things that should be automated but have not been yet, things that are not worth automating, and human processes that can't be automated.

Tasks That Are Not Yet Automated

It takes time to create, test, and deploy automation, so there will always be things that are waiting to be automated. There is never enough time to automate everything, so we must prioritize and choose our methods wisely.

For processes that are not, or have not yet been, automated, creating procedural documentation, called a **playbook**, helps make the process repeatable and consistent. A good playbook makes it easier to automate the process in the future. Often the most difficult part of automating something is simply describing the process accurately. If a playbook does that, the actual coding is relatively easy.

Tasks That Are Not Worth Automating

Some things are not worth automating because they happen infrequently, they are too difficult to automate, or the process changes so often that automation is not possible. Automation is an investment in time and effort and the return on investment (ROI) does not always make automation viable.

Nevertheless, there are some common cases that are worth automating. Often when those are automated, the more rare cases (**edge cases**) can be consolidated or eliminated. In many situations, the newly automated common case provides such superior service that the edge-case customers will suddenly lose their need to be so unique.

Benefits of Automating the Common Case

At one company there were three ways that virtual machines were being provisioned. All three were manual processes, and customers often waited days until a system administrator was available to do the task. A project to automate provisioning was stalled because of the complexity of handling all three variations. Users of the two less common cases demanded that their provisioning process be different because they were (in their own eyes) unique and beautiful snowflakes. They had very serious justifications based on very serious (anecdotal) evidence and waved their hands vigorously to prove their point. To get the project moving, it was decided to automate just the most common case and promise the two edge cases would be added later.

This was much easier to implement than the original all-singing, all-dancing, provisioning system. With the initial automation, provisioning time was reduced to a few minutes and could happen without system administrator involvement. Provisioning could even happen at night and on weekends. At that point an amazing thing happened. The other two cases suddenly discovered that their uniqueness had vanished! They adopted the automated method. The system administrators never automated the two edge cases and the provisioning system remained uncomplicated and easy to maintain.

Tasks That Cannot Be Automated

Some tasks cannot be automated because they are human processes: maintaining your relationship with a stakeholder, managing the bidding process to make a large purchase, evaluating new technology, or negotiating within a team to assemble an oncall schedule. While they cannot be eliminated through automation, they can be streamlined:

- Many interactions with stakeholders can be eliminated through better documentation. Stakeholders can be more self-sufficient if provided with introductory documentation, user documentation, best practices recommendations, a style guide, and so on. If your service will be used by many other services or service teams, it becomes more important to have good documentation. Video instruction is also useful and does not require much effort if you simply make a video recording of presentations you already give.
- Some interactions with stakeholders can be eliminated by making common requests self-service. Rather than meeting individually with customers to understand future capacity requirements, their forecasts can be collected via a web user interface or an API. For example, if you provide a service to hundreds of other teams, forecasting can be become a full-time job for a project manager; alternatively, it can be very little work with proper

automation that integrates with the company's supply-chain management system.

- Evaluating new technology can be labor intensive, but if a common case is identified, the end-to-end process can be turned into an assembly-line process and optimized. For example, if hard drives are purchased by the thousand, it is wise to add a new model to the mix only periodically and only after a thorough evaluation. The evaluation process should be standardized and automated, and results stored automatically for analysis.

- Automation can replace or accelerate team processes. Creating the oncall schedule can evolve into a chaotic mess of negotiations between team members battling to take time off during an important holiday. Automation turns this into a self-service system that permits people to list their availability and that churns out an optimal schedule for the next few months. Thus, it solves the problem better and reduces stress.

- Meta-processes such as communication, status, and process tracking can be facilitated through online systems. As teams grow, just tracking the interaction and communication among all parties can become a burden. Automating that can eliminate hours of manual work for each person. For example, a web-based system that lets people see the status of their order as it works its way through approval processes eliminates the need for status reports, leaving people to deal with just exceptions and problems. If a process has many complex handoffs between teams, a system that provides a status dashboard and automatically notifies teams when hand-offs happen can reduce the need for legions of project managers.

- The best process optimization is elimination. A task that is eliminated does not need to be performed or maintained, nor will it have bugs or security flaws. For example, if production machines run three different operating systems, narrowing that number down to two eliminates a lot of work. If you provide a service to other service teams and require a lengthy approval process for each new team, it may be better to streamline the approval process by automatically approving certain kinds of users.

12.2 Service Life Cycle

Operations is responsible for the entire **service life cycle**: launch, maintenance (both regular and emergency), upgrades, and decommissioning. Each phase has unique requirements, so you'll need a strategy for managing each phase differently.

The stages of the life cycle are:

- **Service Launch**: Launching a service the first time. The service is brought to life, initial customers use it, and problems that were not discovered prior to the launch are discovered and remedied.
- **Emergency Tasks**: Handling exceptional or unexpected events. This includes handling outages and, more importantly, detecting and fixing conditions that precipitate outages.
- **Nonemergency Tasks**: Performing all manual work required as part of the normally functioning system. This may include periodic (weekly or monthly) maintenance tasks (for example, preparation for monthly billing events) as well as processing requests from users (for example, requests to enable the service for use by another internal service or team).
- **Upgrades**: Deploying new software releases and hardware platforms. The better we can do this, the more aggressively the company can try new things and innovate. Each new software release is built and tested before deployment. Tests include system tests, done by developers, as well as user acceptance tests (UAT), done by operations. UAT might include tests to verify there are no **performance regressions** (unexpected declines in performance). Vulnerability assessments are done to detect security issues. New hardware must go through a **hardware qualification** to test for compatibility, performance regressions, and any changes in operational processes.
- **Decommissioning**: Turning off a service. It is the opposite of a service launch: removing the remaining users, turning off the service, removing references to the service from any related service configurations, giving back any resources, archiving old data, and erasing or scrubbing data from any hardware before it is repurposed, sold, or disposed.
- **Project Work**: Performing tasks large enough to require the allocation of dedicated resources and planning. While not directly part of the service life cycle, along the way tasks will arise that are larger than others. Examples include fixing a repeating but intermittent failure, working with stakeholders on roadmaps and plans for the product's future, moving the service to a new datacenter, and scaling the service in new ways.

Most of the life-cycle stages listed here are covered in detail elsewhere in this book. Service launches and decommissioning are covered in detail next.

12.2.1 Service Launches

Nothing is more embarrassing than the failed public launch of a new service.

Often we see a new service launch that is so successful that it receives too much traffic, becomes overloaded, and goes down. This is ironic but not funny.

Each time we launch a new service, we learn something new. If we launch new services rarely, then remembering those lessons until the next launch is difficult. Therefore, if launches are rare, we should maintain a checklist of things to do and record the things you should remember to do next time. As the checklist grows with each launch, we become better at launching services.

If we launch new services frequently, then there are probably many people doing the launches. Some will be less experienced than others. In this case we should maintain a checklist to share our experience. Every addition increases our **organizational memory**, the collection of knowledge within our organization, thereby making the organization smarter.

A common problem is that other teams may not realize that planning a launch requires effort. They may not allocate time for this effort and surprise operations teams at or near the launch date. These teams are unaware of all the potential pitfalls and problems that the checklist is intended to prevent. For this reason the launch checklist should be something mentioned frequently in documentation, socialized among product managers, and made easy to access. The best-case scenario occurs when a service team comes to operations wishing to launch something and has been using the checklist as a guide throughout development. Such a team has "done their homework"; they have been working on the items in the checklist in parallel as the product was being developed. This does not happen by accident; the checklist must be available, be advertised, and become part of the company culture.

A simple strategy is to create a checklist of actions that need to be completed prior to launch. A more sophisticated strategy is for the checklist to be a series of questions that are audited by a Launch Readiness Engineer (LRE) or a Launch Committee.

Because a launch is complex, with many moving parts, we recommend that a single person (the **launch lead**) take a leadership or coordinator role. If the developer and operations teams are very separate, one person from each might be selected to represent each team.

The launch lead then works through the checklist, delegating work, filing bugs for any omissions, and tracking all issues until launch is approved and executed. The launch lead may also be responsible for coordinating post-launch problem resolution.

Case Study: Self-Service Launches at Google

Google launches so many services that it needed a way to make the launch process streamlined and able to be initiated independently by a team. In addition to providing APIs and portals for the technical parts, the Launch Readiness Review (LRR) made the launch process itself self-service.

The LRR included a checklist and instructions on how to achieve each item. An SRE engineer was assigned to shepherd the team through the process and hold them to some very high standards.

Some checklist items were technical—for example, making sure that the Google load balancing system was used properly. Other items were cautionary, to prevent a launch team from repeating other teams' past mistakes. For example, one team had a failed launch because it received 10 times more users than expected. There was no plan for how to handle this situation. The LRR checklist required teams to create a plan to handle this situation and demonstrate that it had been tested ahead of time.

Other checklist items were business related. Marketing, legal, and other departments were required to sign off on the launch. Each department had its own checklist. The SRE team made the service visible externally only after verifying that all of those sign-offs were complete.

12.2.2 Service Decommissioning

Decommissioning (or just "decomm"), or turning off a service, involves three major phases: removal of users, deallocation of resources, and disposal of resources.

Removing users is often a product management task. Usually it involves making the users aware that they must move. Sometimes it is a technical issue of moving them to another service. User data may need to be moved or archived.

Resource deallocation can cover many aspects. There may be DNS entries to be removed, machines to power off, database connections to be disabled, and so on. Usually there are complex dependencies involved. Often nothing can begin until the last user is off the service; certain resources cannot be deallocated before others, and so on. For example, typically a DNS entry is not removed until the machine is no longer in use. Network connections must remain in place if deallocating other services depends on network connectivity.

Resource disposal includes securely erasing disks and other media and disposing of all hardware. The hardware may be repurposed, sold, or scrapped.

If decommissioning is done incorrectly or items are missed, resources will remain allocated. A checklist, that is added to over time, will help assure decommissioning is done completely and the tasks are done in the right order.

12.3 Organizing Strategy for Operational Teams

An operational team needs to get work done. Therefore teams need a strategy that assures that all incoming work is received, scheduled, and completed. Broadly speaking, there are three sources of operational work and these work items fall into three categories. To understand how to best organize a team, first you must understand these sources and categories.

The three sources of work are life-cycle management, interacting with stakeholders, and process improvement and automation. Life-cycle management is the operational work involved in running the service. Interacting with stakeholders refers to both maintaining the relationship with people who use and depend on the service, and prioritizing and fulfilling their requests. Process improvement and automation is work inspired by the business desire for continuous improvement.

No matter the source, this work tends to fall into one of these three broad categories:

- **Emergency Issues:** Outages, and issues that indicate a pending outage that can be prevented, and emergency requests from other teams. Usually initiated by an alert sent by the monitoring system via SMS or pager.
- **Normal Requests:** Process work (repeatable processes that have not yet been automated), non-urgent trouble reports, informational questions, and initial consulting that results in larger projects. Usually initiated by a request ticket system.
- **Project Work:** Small and large projects that evolve the system. Managed with whatever project management style the team selects.

To assure that all sources and categories of work receive attention, we recommend this simple organizing principle: people should always be working on projects, with exceptions made to assure that emergency issues receive immediate attention and non-project customer requests are triaged and worked in a timely manner.

More specifically, at any given moment, the highest priority for one person on the team should be responding to emergencies, the highest priority for one other person on the team should be responding to normal requests, and the rest of the team should be focused on project work.

This is counter to the way operations teams often work: everyone running from emergency to emergency with no time for project work. If there is no effort dedicated to improving the situation, the team will simply run from emergency to emergency until they are burned out.

Major improvements come from project work. Project work requires concentration and focus. If you are constantly being interrupted with emergency issues and requests, you will not be able to get projects done. If an entire team is focused on emergencies and requests, nobody is working on projects.

It can be tempting to organize an operations team into three subteams, each focusing on one source of work or one category of work. Either of these approaches will create silos of responsibility. Process improvement is best done by the people involved in the process, not by observers.

To implement our recommended strategy, all members of the team focus on project work as their main priority. However, team members take turns being responsible for emergency issues as they arise. This responsibility is called **oncall**. Likewise, team members take turns being responsible for normal requests from other teams. This responsibility is called **ticket duty**.

It is common that oncall duty and ticket duty are scheduled in a rotation. For example, a team of eight people may use an eight-week cycle. Each person is assigned a week where he or she is on call: expected to respond to alerts, spending any remaining time on projects. Each person is also assigned a different week where he or she is on ticket duty: expected to focus on triaging and responding to request tickets first, working on other projects only if there is remaining time. This gives team members six weeks out of the cycle that can be focused on project work.

Limiting each rotation to a specific person makes for smoother handoffs to the next shift. In such a case, there are two people doing the handoff rather than a large operations team meeting. If more than 25 percent of a team needs to be dedicated to ticket duty and oncall, there is a serious problem with firefighting and a lack of automation.

The team manager should be part of the operational rotation. This practice ensures the manager is aware of the operational load and firefighting that goes on. It also ensures that nontechnical managers don't accidentally get hired into the operations organization.

Teams may combine oncall and ticket duty into one position if the amount of work in those categories is sufficiently small. Some teams may need to designate multiple people to fill each role.

Project work is best done in small teams. Solo projects can damage a team by making members feel disconnected or by permitting individuals to work without constructive feedback. Designs are better with at least some peer review. Without feedback, members may end up working on projects they feel are important but have marginal benefit. Conversely, large teams often get stalled by lack of consensus. In their case, focusing on shipping quickly overcomes many of these problems. It helps by making progress visible to the project members, the wider team, and management. Course corrections are easier to make when feedback is frequent.

The Agile methodology is an effective way to organize project work.

There is also meta-work: meetings, status reports, company functions. These generally eat into project time and should be minimized.

12.3.1　Team Member Day Types

Now that we have established an organizing principle for the team's work, each team member can organize his or her work based on what kind of day it is: a project-focused day, an oncall day, or a ticket duty day.

Project-Focused Days

Most days should be project days for operational staff. Specifically, most days should be spent developing software that automates or optimizes aspects of the team's responsibilities. Non-software projects include shepherding a new launch or working with stakeholders on requirements for future releases.

Organizing the work of a team through a single bug tracking system has the benefit of reducing time spent checking different systems for status. Bug tracking systems provide an easy way for people to prioritize and track their work. On a typical project day the staff member starts by checking the bug tracking system to review the bugs assigned to him or her, or possibly to review unassigned issues of higher priority the team member might need to take on.

Software development in operations tends to mirror the Agile methodology: rather than making large, sudden changes, many small projects evolve the system over time.

Projects that do not involve software development may involve technical work. Moving a service to a new datacenter is highly technical work that cannot be automated because it happens infrequently.

Operations staff tend not to physically touch hardware not just because of the heavy use of virtual machines, but also because even physical machines are located in datacenters that are located far away. Datacenter technicians act as **remote hands**, applying physical changes when needed.

Oncall Days

Oncall days are spent working on projects until an alert is received, usually by SMS, text message, or pager.

Once an alert is received, the issue is worked until it is resolved. Often there are multiple solutions to a problem, usually including one that will fix the problem quickly but temporarily and others that are long-term fixes. Generally the quick fix is employed because returning the service to normal operating parameters is paramount.

Once the alert is resolved, a number of other tasks should always be done. The alert should be categorized and annotated in some form of electronic alert journal so that trends may be discovered. If a quick fix was employed, a bug should be filed requesting a longer-term fix. The oncall person may take some time to update the playbook entry for this alert, thereby building organizational memory. If there was a user-visible outage or an SLA violation, a postmortem report should be written. An investigation should be conducted to ascertain the root cause of the problem. Writing a postmortem report, filing bugs, and root causes identification are all ways that we raise the visibility of issues so that they get attention. Otherwise, we will continually muddle through ad hoc workarounds and nothing will ever get better. Postmortem reports (possibly redacted for technical content) can be shared with the user community to build confidence in the service.

The benefit of having a specific person assigned to oncall duty at any given time is that it enables the rest of the team to remain focused on project work. Studies have found that the key to software developer productivity is to have long periods of uninterrupted time. That said, if a major crisis appears, the oncall person will pull people away from their projects to assist.

If oncall shifts are too long, the oncall person will be overloaded with follow-up work. If the shifts are too close together, there will not be time to complete the follow-up work. Many great ideas for new projects and improvements are first imagined while servicing alerts. Between oncall shifts people should have enough time to pursue such projects.

Ticket Duty Days

Ticket duty days are spent working on requests from customers. Here the customers are the internal users of the service, such as other service teams that use your service's API. These are not tickets from external users. Those items should be handled by customer support representatives.

While oncall is expected to have very fast reaction time, tickets generally have an expected response time measured in days.

Typical tickets may consist of questions about the service, which can lead to some consulting on how to use the service. They may also be requests for activation of a service, reports of problems or difficulties people are experiencing, and so forth. Sometimes tickets are created by automated systems. For example, a monitoring system may detect a situation that is not so urgent that it needs immediate response and may open a ticket instead.

Some long-running tickets left from the previous shift may need follow-up. Often there is a policy that if we are waiting for a reply from the customer, every three days the customer will be politely "poked" to make sure the issue is not forgotten. If the customer is waiting for follow-up from us, there may be a policy that urgent tickets will have a status update posted daily, with longer stretches of time for other priorities.

If a ticket will not be completed by the end of a shift, its status should be included in the shift report so that the next person can pick up where the previous person left off.

By dedicating a person to ticket duty, that individual can be more focused while responding to tickets. All tickets can be triaged and prioritized. There is more time to categorize tickets so that trends can be spotted. Efficiencies can be realized by batching up similar tickets to be done in a row. More importantly, by dedicating a person to tickets, that individual should have time to go deeper into each ticket: to update documentation and playbooks along the way, to deep-dive into bugs rather than find superficial workarounds, to fix complex broken processes. Ticket duty should not be a chore, but rather should be part of the strategy to reduce the overall work faced by the team.

Every operations team should have a goal of eliminating the need for people to open tickets with them, similar to how there should always be a goal to automate manual processes. A ticket requesting information is an indication that documentation should be improved. It is best to respond to the question by adding the requested information to the service's FAQ or other user documentation and then directing the user to that document. Requests for service activation, allocations, or configuration changes indicate an opportunity to create a web-based portal or API to make such requests obsolete. Any ticket created by an automated system should have a corresponding playbook entry that explains how to process it, with a link to the bug ID requesting that the automation be improved to eliminate the need to open such tickets.

At the end of oncall and ticket duty shifts, it is common for the person to email out a shift report to the entire team. This report should mention any trends noticed and any advice or status information to be passed on to the next person. The oncall end-of-shift report should also include a log of which alerts were received and what was done in response.

When you are oncall or doing ticket duty, that is your main project. Other project work that is accomplished, if any, is a bonus. Management should not expect other projects to get done, nor should people be penalized for having the proper focus. When people end their oncall or ticket duty time, they should not complain that they weren't able to get any project work done; their project, so to speak, was ticket duty.

12.3.2 Other Strategies

There are many other ways to organize the work of a team. The team can rotate though projects focused on a particular goal or subsystem, it can focus on reducing toil, or special days can be set aside for reducing technical debt.

Focus or Theme

One can pick a category of issues to focus on for a month or two, changing themes periodically or when the current theme is complete. For example, at the start of a theme, a number of security-related issues can be selected and everyone commit to focusing on them until they are complete. Once these items are complete, the next theme begins. Some common themes include monitoring, a particular service or subservice, or automating a particular task.

If the team cohesion was low, this can help everyone feel as if they are working as a team again. It can also enhance productivity: if everyone has familiarized themselves with the same part of the code base, everyone can do a better job of helping each other.

Introducing a theme can also provide a certain amount of motivation. If the team is looking forward to the next theme (because it is more interesting, novel, or fun), they will be motivated to meet the goals of the current theme so they can start the next one.

Toil Reduction

Toil is manual work that is particularly exhausting. If a team calculates the number of hours spent on toil versus normal project work, that ratio should be as low as possible. Management may set a threshold such that if it goes above 50 percent, the team pauses all new features and works to solve the big problems that are the source of so much toil. (See Section 12.4.2.)

Fix-It Days

A day (or series of days) can be set aside to reduce technical debt. **Technical debt** is the accumulation of small unfinished amounts of work. By themselves, these bits and pieces are not urgent, but the accumulation of them starts to become a problem. For example, a Documentation Fix-It Day would involve everyone stopping all other work to focus on bugs related to documentation that needs to be improved. Alternatively, a Fix-It Week might be declared to focus on bringing all monitoring configurations up to a particular standard.

Often teams turn fix-its into a game. For example, at the start a list of tasks (or bugs) is published. Prizes are given out to the people who resolve the most bugs. If done company-wide, teams may receive T-shirts for participating and/or prizes for completing the most tasks.

12.4 Virtual Office

Many operations teams work from home rather than an office. Since work is virtual, with remote hands touching hardware when needed, we can work from anywhere. Therefore, it is common to work from anywhere. When necessary, the team meets in chat rooms or other virtual meeting spaces rather than physical meeting rooms. When teams work this way, communication must be more intentional because you don't just happen to see each other in the office.

It is good to have a policy that anyone who is not working from the office takes responsibility for staying in touch with the team. They should clearly and periodically communicate their status. In turn, the entire team should take responsibility for making sure remote workers do not feel isolated. Everyone should know what their team members are working on and take the time to include everyone in discussions. There are many tools that can help achieve this.

12.4.1 Communication Mechanisms

Chat rooms are commonly used for staying in touch throughout the day. Chat room transcripts should be stored and accessible so people can read what they may have missed. There are many chat room "bots" (software robots that join the chat room and provide services) that can provide transcription services, pass messages to offline members, announce when oncall shifts change, and broadcast any alerts generated by the monitoring system. Some bots provide entertainment: At Google, a bot keeps track of who has received the most virtual high-fives. At Stack Exchange, a bot notices if anyone types the phrase "not my fault" and responds by selecting a random person from the room and announcing this person has been randomly designated to be blamed.

Higher-bandwidth communication systems include voice and video systems as well as screen sharing applications. The higher the bandwidth, the better the fidelity of communication that can be achieved. Text-chat is not good at conveying emotions, while voice and video can. Always switch to higher-fidelity communication systems when conveying emotions is more important, especially when an intense or heated debate starts.

The communication medium with the highest fidelity is the in-person meeting. Virtual teams greatly benefit from periodic in-person meetings. Everyone travels to the same place for a few days of meetings that focus on long-term planning, team building, and other issues that cannot be solved online.

12.4.2 Communication Policies

Many teams establish a communication agreement that clarifies which methods will be used in which situations. For example, a common agreement is that chat rooms will be the primary communication channel but only for ephemeral discussions. If a decision is made in the chat room or an announcement needs to be made, it will be broadcast via email. Email is for information that needs to carry across oncall shifts or day boundaries. Announcements with lasting effects, such as major policies or design decisions, need to be recorded in the team wiki or other document system (and the creation of said document needs to be announced via email). Establishing this chat - email - document paradigm can go a long way in reducing communication problems.

12.5 Summary

Operations is different from typical enterprise IT because it is focused on a particular service or group of services and because it has more demanding uptime requirements.

There is a tension between the operations team's desire for stability and the developers' desire to get new code into production. There are many ways to reach a balance. Most ways involve aligning goals by sharing responsibility for both uptime and velocity of new features.

Operations in distributed computing is done at a large scale. Processes that have to be done manually do not scale. Constant process improvement and automation are essential.

Operations is responsible for the life cycle of a service: launch, maintenance, upgrades, and decommissioning. Maintenance tasks include emergency and non-emergency response. In addition, related projects maintain and evolve the service.

Launches, decommissioning of services, and other tasks that are done infrequently require an attention to detail that is best assured by use of checklists. Checklists ensure that lessons learned in the past are carried forward.

The most productive use of time for operational staff is time spent automating and optimizing processes. This should be their primary responsibility. In addition, two other kinds of work require attention. Emergency tasks need fast response. Nonemergency requests need to be managed such that they are prioritized and worked in a timely manner. To make sure all these things happen, at any given time one person on the operations team should be focused on responding to emergencies; another should be assigned to prioritizing and working on nonemergency requests. When team members take turns addressing these responsibilities, they receive the dedicated resources required to assure they happen correctly by sharing the responsibility across the team. People also avoid burning out.

Operations teams generally work far from the actual machines that run their services. Since they operate the service remotely, they can work from anywhere there is a network connection. Therefore teams often work from different places, collaborating and communicating in a chat room or other virtual office. Many tools are available to enable this type of organizational structure. In such an environment, it becomes important to change the communication medium based on the type of communication required. Chat rooms are sufficient for general communication but voice and video are more appropriate for more intense discussions. Email is more appropriate when a record of the communication is required, or if it is important to reach people who are not currently online.

12.6 New Words and Expressions

SRE (Site Reliability Engineering)	网站可靠性工程
Stakeholder	*n.* 利益相关者
Priority	*n.* 优先顺序
outage	*n.* 中断供应；运行中断
SLA (Service Level Agreement)	*abbr.* 服务水平协议
bespoke system	定制系统
off the shelf	现成的，不用定制的
scalability	*n.* 可扩展性；可伸缩性
tangible	*adj.* 有形的；切实的
viable	*adj.* 可行的；可实行的
ROI (return on investment)	*abbr.* 投资回收率
anecdotal	*adj.* 轶事的；多轶事的
vigorously	*adv.* 精神旺盛地，活泼地
regression	*n.* 回归；退化
potential pitfall	潜在危险
DNS (Domain Name Server)	*abbr.* 域名服务器
emergency	*n.* 紧急情况；突发事件
pending	*adj.* 马上就要发生的
SMS (Short Messaging Service)	*abbr.* 短信服务
optimize	*vt.* 使最优化，使完善
shepherd	*vt.* 带领；指导
bug tracking systems	错误跟踪系统
paramount	*adj.* 最重要的，主要的
postmortem	*n.* 事后析误
ascertain	*vt.* 确定；查明
user community	用户社区
and so forth	等等

immediate response 立即反应
triage vt. 鉴别分类
FAQ (Frequently Asked Question) abbr. 常见问题
toil n. 辛苦；苦工
threshold n. 入口；门槛；临界值
fidelity n. 保真度；精确
boundary n. 边界，界线
announcement n. 公告；通告

12.7 Questions

Single Choice Questions

1. Which of the following does not represent the difference between the SRE and Traditional Enterprise IT? (   )

　　A. A single service or a well-defined set of services

　　B. The attitude toward uptime

　　C. They are not "off the shelf" or turn key systems

　　D. Efficiency is important

2. Which of the following terms best describes the tension between the desire for stability and the desire for change? (   )

　　A. Operations teams don't tend to favor stability; developers desire change

　　B. Operations teams tend to favor stability; developers don't desire change

　　C. Operations teams tend to favor stability; developers desire change

　　D. Operations teams don't tend to favor stability; developers don't desire change

3. Oncall teams are made up of different members at least at one or two locations. Which is the right number of the following? (   )

　　A. Six members at one location, or eight members at two locations

　　B. Eight members at one location, or six members at two locations

　　C. Eight members at one location, or eight members at two locations

　　D. Six members at one location, or six members at two locations

4. Three categories of things are not automated except which of the following? (   )

　　A. Things that should be automated but have not been yet

　　B. Things that are not worth automating

　　C. Human processes that are bored

　　D. Human processes that can't be automated

5. Which of the following represents the three major phases of decommissioning

or turning off a service? (   )

　　A. Removal of users, deallocation of resources, and disposal of resources

　　B. Removal of users, location of resources, and disposal of resources

　　C. Removal of users, deallocation of resources, and proposal of resources

　　D. Removal of users, location of resources, and proposal of resources

6. To best organize a team tends to fall into one of these three broad categories: (   ).

　　A. Emergency Requests, Normal Issues and Project Work

　　B. Emergency Issues, Normal Requests and Project Work

　　C. Emergency Work, Normal Requests and Project Issues

　　D. Emergency Issues, Normal Work and Project Requests

7. To assure that all sources and categories of work receive attention, which of the following simple organizing principle is false at any given moment? (   )

　　A. The highest priority for one person on the team should be responding to emergencies

　　B. The highest priority for one other person on the team should be responding to normal requests

　　C. The rest of the team should be focused on project work

　　D. None of the above

8. What kind of requests for service changes can not indicate an opportunity to create a web-based portal or API to make such requests obsolete? (   )

　　A. activation　　　　　　　　B. allocations

　　C. concentration　　　　　　 D. configuration

9. Which of the following statements is true? (   )

　　A. Text-chat is not good at conveying emotions, while voice and video can

　　B. Voice-chat is not good at conveying emotions, while text and video can

　　C. Video-chat is not good at conveying emotions, while text and voice can

　　D. Text-chat is good at conveying emotions, while voice and video can

10. Which kind of the following communication channel is ephemeral? (   )

　　A. Email　　　　　　　　　　B. Chat room

　　C. Wiki　　　　　　　　　　 D. Document system

12.8　Problems

After reading this chapter and completing the exercises, you will be able to do the following:

1. What is operations? What are its major areas of responsibilities?

2. How does operations in distributed computing differ from traditional desktop

support or enterprise client—server support?

3. Describe the service life cycle as it relates to a service you have experience with.

4. Section 12.1.2 discusses the change-instability cycle. Draw a series of graphs where the x-axis is time and the y-axis is the measure of stability. Each graph should represent two months of project time.

Each Monday, a major software release that introduces instability (9 bugs) is rolled out. On Tuesday through Friday, the team has an opportunity to roll out a "bug-fix" release, each of which fixes three bugs. Graph these scenarios:

   a. No bug-fix releases
   b. Two bug-fix releases after every major release
   c. Three bug-fix releases after every major release
   d. Four bug-fix releases after every major release
   e. No bug-fix release after odd releases, five bug-fix releases after even releases

5. What do you observe about the graphs from Exercise 4?

6. For a service you provide or have experience with, who are the stakeholders? Which interactions did you or your team have with them?

7. What are some of the ways operations work can be organized? How does this compare to how your current team is organized?

8. For a service you are involved with, give examples of work whose source is life-cycle management, interacting with stakeholders, and process improvement and automation.

9. For a service you are involved with, give examples of emergency issues, normal requests, and project work.

Answers to Questions

Unit 1

1. D 2. A 3. A 4. C 5. A 6. B 7. D 8. D 9. B 10. D

Unit 2

1. D 2. C 3. B 4. D 5. B 6. C 7. B 8. A 9. D 10. A

Unit 3

1. C 2. D 3. B 4. B 5. B 6. D 7. A 8. C 9. A 10. C

Unit 4

1. B 2. C 3. A 4. D 5. A 6. B 7. B 8. C 9. A 10. D

Unit 5

1. C 2. A 3. B 4. D 5. D 6. B 7. D 8. C 9. A 10. B

Unit 6

1. A 2. B 3. C 4. D 5. A 6. B 7. C 8. B 9. D 10. B

Unit 7

1. C 2. A 3. D 4. B 5. A 6. D 7. A 8. A 9. B 10. C

Unit 8

1. C 2. D 3. B 4. D 5. A 6. B 7. D 8. D 9. C 10. B

Unit 9

1. B 2. A 3. C 4. C 5. C 6. D 7. D 8. C 9. B 10. D

Unit 10

1. D 2. B 3. C 4. A 5. D 6. A 7. D 8. C 9. B 10. D

Unit 11

1. A 2. B 3. A 4. C 5. D 6. B 7. D 8. A 9. C 10. B

Unit 12

1. D 2. C 3. B 4. C 5. A 6. B 7. D 8. C 9. A 10. B

References

1. T. S. Lavergetta. Microwave Materials and Fabrication Techniques. Dedham. MA: Artech House, Inc. ,1984.
2. K. C. Gupta. Microstrip Lines and Slot Lines. 2nd ed. Dedham. MA: Artech House, Inc. ,1996.
3. Kyung-Whan Yeom. Microwave Circuit Design: A Practical Approach Using ADS. Prentice Hall. May 22,2015.
4. Scott Daley. Project 2013 In Depth. Que, Aug 23, 2013.
5. Pramod J. Sadalage, Martin Fowler. NoSQL Distilled: A Brief Guide to the Emerging World of Polyglot Persistence. Addison-Wesley Professional Aug 8, 2012.
6. Len Bass, Ingo Weber, Liming Zhu. DevOps: A Software Architect's Perspective. Addison-Wesley Professional. May 18, 2015.
7. Stephen O'Brien, Ultimate Player's Guide to Minecraft - Xbox Edition. The: Covers both Xbox 360 and Xbox One Versions. Que, Nov 17,2014.
8. Amir Ranjbar. Troubleshooting and Maintaining Cisco IP Networks (TSHOOT) Foundation Learning Guide: (CCNP TSHOOT 300-135). Cisco Press. Dec 31, 2014.
9. Daniel Knott. Hands-On Mobile App Testing: A Guide for Mobile Testers and Anyone Involved in the Mobile App Business. Addison-Wesley Professional, May 18, 2015.
10. Tim Hayden, Tom Webster. Mobile Commerce Revolution. The: Business Success in a Wireless World, Que, Oct 1,2014.
11. Pramod J. Sadalage, Martin Fowler. NoSQL Distilled: A Brief Guide to the Emerging World of Polyglot Persistence. Addison-Wesley Professional, Aug 8, 2012.
12. Philip Russom. Big Data Analytics. TDWI Research Center Director for Data Management, October 11,2011.
13. Peter Wayner. 7 top tools for taming big data. Follow InfoWorld, Apr 18,2012.
14. Thomas A. Limoncelli, Strata R. Chalup, Christina J. Hogan. Practice of Cloud System Administration. The: Designing and Operating Large Distributed Systems, Volume 2, Addison-Wesley Professional, Sep 3,2014.
15. Mark S. Merkow, Jim Breithaupt. Information Security: Principles and Practices. 2nd Edition. Pearson IT Ceritication, Jun 4,2014.